WHY THE UNIVERSE BOTHERS TO EXIST

Theistic Determinism, Evidences and Implications—A Worldview Proposal

By
David V. McCorkle

Strategic Book Publishing and Rights Co.

Strategic Book Publishing and Rights Co.
12620 FM 1960, Suite A4-507
Houston, TX 77065
www.sbpra.com

ISBN: 978-1- 62212-580-7

Design: Dedicated Book Services, (www.netdbs.com)

TABLE OF CONTENTS

This book is dedicated to
College students
And all others who seek truth

Prologue

The primary objective of this book is to establish a resolution to the origins controversy that has pitted biblical perspective against science for too long. It is now established that our universe is amazingly fine-tuned, allowing our just-right planet to harbor life forms including us as sentient beings. I argue that this fine-tuning extends to the conditions that allow a dynamic fit of populations of organisms to changes in their environment. Chronological modification of species accommodates econiche dynamics so that the relationships remain fine-tuned. If one attributes the fine-tuning of the universe to God's design, in that God is immanent (omnipresent) in the actualization of this design, why should God not also be immanent in the dynamics of living systems? That God is so involved is certainly supported biblically. (For instance, see Psalm 104.)

This book's preface attempts to set a context for its perspective on existence. It involves what has been a challenge for thinkers over the ages. Here I do not attempt to document this history, but rather try only to set the conceptual stage for the theme of the book: why the universe and, indeed, anything at all, exists. (But that it does exist raises the question: Can there be anything that does not exist including God?) For those who wish to pursue the historical aspects of attempts to deal with this dilemma, there are other works that treat this history well. For instance, Frank Close in his book *Nothing, a Very Short Introduction*, Oxford University Press 2009, gives a survey of the history of perspectives on the concept of nothingness. (Apparently, the Greek philosopher Thales agreed with my perspective back in 600 BC!) Close's book also gives references for further pursuit of the subject.

A book authored by Henning Genz titled *Nothingness, the Science of Empty Space* (Perseus Publishing, Cambridge Mass., English translation, 1999), might also be of interest to the reader wishing to expand on the subject, at least as it deals with what I consider the "relative nothingness" within our universe. Genz also goes in depth into the history of the subject.

In that my book deals with several specialty areas, expect the use of the respective vocabularies. Referring to a dictionary, or e.g., Google on the internet, may be of help to readers willing to learn appropriately if such terminologies are unfamiliar.

I take special note of the implications of Steinhardt and Turok's proposed ekpyrotic universe model. If, indeed, this concept is accurate, heaven may not be far away! I suggest that the biblical Garden of Eden still exists but in the universe opposed to ours in this model. God places a man taken from the earthly population of *Homo sapiens* into the Garden where the events of Genesis 2 take place. Mature sentience is realized when Eve and Adam partake of the forbidden fruit of the tree of the "knowledge of good and evil." Upon their banishment back to planet Earth, their progeny intermarry with the remnant population of our species. Thus, Adam and Eve were, indeed, the first fully sentient progenitors of modern humans. Our species, however, originated as a population, as anthropology documents.

It is also of special note that aspects of the Genesis 1 account amazingly function as unwitting allegory of the ekpyrotic model itself. But Andrew Parker in his book *The Genesis Enigma: Why the First Book of the Bible Is Scientifically Accurate* (see our bibliography) finds concordance in some of the same Genesis passages with aspects of the scientific account of life's origin and development on our planet. In my mind, that the same ancient statements can document divine inspiration through allegory for more than one scientifically demonstrated reality is nothing short of astounding!

In this book, *Why the Universe Bothers to Exist*, I suggest that at least one means of God's mode of action in nature is by discretionary control of decoherence of probability waves. I first encountered this perspective in my reading of works by Nancy Murphy, et al., Arthur Peacocke, and John Polkinghorne.

Another aspect I treat at length is an account of the unique need among species for a moral base in *Homo sapiens*. I suggest a concept I term "socionics" as foundational to such a need, yet a part of the process of theistic determinism.

I find fulfilment in "pushing intellectual buttons." I truly hope that this book will stimulate you, the reader, to further investigate the ideas and develop concepts of your own to be shared in the controversy forum as it develops. I trust that you will always relate your conclusions to biblical text.

The "new wine" here proposed is perhaps appropriately held in the "new wineskin" of the recognition of what I term "unwitting allegory" in biblical content.

Please drink freely!

David V. McCorkle, PhD Zoology, University of Washington. Professor Emeritus, Western Oregon University.

Introduction:

Why Read This Book?

This book proposes a resolution to the origins controversy that is both scientific and in concordance with biblical scriptures. It holds that humans, as a species, originated as a population (Genesis one), but as sentient beings from one pair, Adam and Eve, who existed in a parallel universe (ekpyrotic model, "east of Eden"), until the fall (Genesis two). It also provides answers to several other perplexing questions faced by theists such as the following:

What philosophical explanation is there for the existence of somethingness including God and our universe?

Is somethingness infinite or does it have boundary?

What is the evidence that God exists? What is God like?

Where is heaven? What will we do in heaven? Why will a new heaven eventually be needed?

Why will "redeemed" humans need to be resurrected if they are already in heaven?

What happened to the Garden of Eden? Might it still exist? If so, where?

How could information of our personal identity be preserved after our physical death?

What determines which potential realities actualize when probability waves collapse into eigenstates?

http://sbpra.com/DavidVMcCorkle/

Preface: Establishing a Model: Somethingness versus Nothingness

In the concluding chapter of his book, *A Brief History of Time*, Stephen Hawking poses a fundamental question: Why does the universe bother to exist?[1] As fundamental a question as this is, its corollary, why does somethingness exist at all, is perhaps the most basic of all questions. How then can such questions be answered? To what degree empirically and where might logic lead on the matter? One answer: Somethingness cannot exist and therefore does not, can serve as a null hypothesis, and is falsified empirically by even our own existence. But our form of somethingness is relative and apparently finite. What of "absolute, transcendent somethingness"?[2]

Perhaps absolute somethingness exists because it cannot not exist. If absolute nothingness (as distinct from relative nothingness), cannot be, then how could absolute somethingness not be? (If absolute nothingness does exist, then, by definition, it must be some sort of somethingness.) If our logic holds to this point, we must then conclude that absolute somethingness is infinite. It could not have a boundary with absolute nothingness if the latter is non-existent. Nor could it have an origin, which implies a boundary in "time."[3]

The conclusion that absolute somethingness is infinite does not, however, necessitate that relative somethingnesses are infinite. They could obviously border with other relative somethingness (which could, perhaps, be thought of as relative nothingness). Thus, the somethingness of our universe may not be infinite.[4]

Now if infinite, absolute, somethingness has no existence boundary,[5] does this include no limit to its complexity? Can there, in fact, even be a boundary to its simplicity? In either case, such a boundary would have to be with absolute, non-existent nothingness, which we hold to be logically impossible. So perhaps in its simplicity it is continuous in the form of an infinite negative somethingness, perhaps even increasing in "negative complexity," whatever that may mean.

And what, then, could be a limit to the complexity of absolute somethingness? And if such complexity is infinite, how could it not include sentience (conscious awareness)? Obviously, we are here to ask questions, evidencing the existence of sentience at least in our version of relative somethingness. And could not our sentience be somehow reflective of an infinite sentience that infinite complexity, transcendent to our relative somethingness, would require? Although our complexity, including our sentience, can be taken to be emergent within our circumstance, nonetheless, its potential could be eternal. (See [3].)

We are here at a critical point in our logic. Does God exist or not? So let us now explore the possibility of empirical evidence for an answer. This brings us to the "is the island inhabited?" dilemma. Does the lack of observing footprints prove empirically that an island is not inhabited? On the other hand, a single footprint, that is any positive evidence, demonstrates empirically that it is. In the case of God's (as here defined) existence, we are unable to examine but a tiny portion of the "island." Thus, to conclude that God does not exist because one hasn't seen his "footprint," that is upon non-comprehensive negative evidence alone, would clearly be illogical.[6] Making no conclusion on the matter at this point (agnosticism), however, is not out of order logically. And to logically conclude that God does exist, that, indeed, infinite, absolute somethingness includes infinite sentience, should depend upon positive evidence. Fortunately, we can seek this even in our small part of the "island." So let us now search for such evidence.

CHAPTER ONE:
Has God Left His Footprint?

To attempt to answer this question, we need first to understand what we are looking for. How is God's footprint to be recognized? That is, what constitutes positive evidence of at least immanent (entanglement in a comprehensive sense) sentience within our universe? The classic example of Paley's watch would give evidence of order imposed from outside the ordered product, implying the involvement of at least finite sentience. Here we look for intrinsic order such as is to be expected of at least immanent, if not transcendent, sentience.[7] Thus, the whole of our practice of science is really a documentation of such evidence[8]. In this perspective, science is the true friend of theists, hardly their enemy. Such intrinsic order, then, implies meaning and purpose (design) if attributed to immanent sentience. Perhaps our existence as sentient, although finite, beings constitutes an outcome of such design. Could we ourselves, then, constitute the footprints we look for?

Thus, as we have seen, the positive evidence of God's presence would be order immanent to somethingness, such order implying intent. It follows, then, for instance, that the Anthropic Principle, indeed, would be the product of design as would the existence of "natural law." But how might such order originate? And what of the uncertainty of quantum mechanics? Ahh! But could not this uncertainty be the "sand" that God's immanent presence "distorts and compresses" creating his footprint of order? The decoherence of quantum probability waves in meaningful patterns would be at least a means of God's action, and thus immanent to our relative somethingness.[9] Indeed, if decoherence does not occur "meaningfully," then how is it that, for instance, our existence is sustained? (True, open systems offset entropy through the input of energy. But here we refer to the basic order that underlies the very existence of such systems. Order emergent from quantum uncertainty, upon which order even entropic states of supraquantum somethingness would depend.) So we might rephrase Stephen Hawking's question thus, "Why does the universe go to all the bother of existing and persisting in its orderliness?" How is it that order presides over the disorder implied by uncertainty?[10]

From the perspective we have developed, it follows that whatever imposes order on the quantum level resulting in the emergence of the ordered complexity characteristic of our universe is recognized as the creating action of infinite sentience. What we recognize as laws that govern this process have primary existence, then, in this infinite sentience, in other words in God's "mind." And the actualized outcomes, that is, what emerges from the process, is reflective of God's intentions.

Note, however, that this is a dynamic process. Quantum uncertainty both denies Newtonian determinism and provides means for God's responsive action. Final outcomes are intended in the eternal God's mind (hence theistic determinism replaces Newtonian determinism). But the actualization of these intended outcomes is dynamic. The rapid regeneration of quantum probability waves allows flexibility in decoherence patterns as God responds to circumstances to the degree that he allows their free running.[11] Thus, his intentions are actualized within a system based upon uncertainty. (Note that this allows us free will as well as God's sovereignty.) God's footprint is imprinted in the sands of quantum uncertainty.

So order and its emergent complexity is the "footprint" of God that we seek. And what better represents this emergent complexity than human sentience? Hence, indeed, we are God's "footprints," reflections of his reality. No wonder, then, that our "impurities" of sentience (sins) are a problem. How are we to be a true reflection of God's image if we are imperfect? Obviously, God must deal with such impurity if his intentions are to be realized. False footprints cannot be tolerated if God is to maintain the integrity of his identity. And here we encounter the problem of "good" versus "bad." What is it that contaminates us as true "footprints"?

Some Thoughts on Goodness versus Badness

Is too much good bad? Yes, if resource is limited as here on Earth. No, if resource is infinite as with God. Thus, God can be infinitely good without being bad.[12]

On Earth there is limited resource and potentially infinite demand. Therefore, some "bad" is good. Relative bad, that is what seems bad to individuals, may act as negative feedback keeping resource demand in balance with supply. Thus, ultimate good is the result. For instance, if the reproductive potential for a population were to become actual and all individuals survived and thrived, actualizing their reproductive potential in turn and so on, finite resource would be exhausted and the system would collapse.[13] In contrast, if there is negative feedback such as by disease, predation, natural disaster, etc. that limits reproduction and survival, then the system may stay in balance so that ultimate good is served. Thus, for God to create relative "badness" is actually good and his infinitely good character is not violated. God may use relative bad in his creative process and yet retain his absolute, good character.

God's goodness, then, is absolute and infinite so long as somethingness (resource) is infinite. And therefore, there can be no absolute badness.[14] (If there were, then goodness could not be infinite.) Thus, badness of this definition is only relative in closed systems with finite resource. However, disobedience to God is a challenge to his infinite goodness. But only sentient beings have the potential to disobey.[15] Thus, "sin" (disobedience) is different from the "badness" in natural systems. God's character would be compromised if he did not rectify such disobedience and thereby purify "contaminated relative sentience" because this condition constitutes a false reflection of his image, a false "footprint." (It would follow that much of human violence and ill will is sin especially when peaceful, alternative negative feedback is an option.)

The Role of Randomness in God's Action

From our finite perspective, we may have a specific objective for our action. For instance, we may wish to control the wetness of our flower bed. If it is too dry, we use a hose to water it, considering water undirected at the flower bed a waste. But what should be expected of infinite sentience actualizing objectives? When God waters with a rainstorm, from his infinite perspective and in role with his immanency, there would be multiple objectives and therefore no waste.[16] This principle has implications. What seems random from a relative perspective, from an infinite perspective, indeed, must have purpose.

If one holds an eyedropper of water over a sheet of paper, squeezing out a drop, its destination on the paper is easily predicted. If held close enough, one would not have difficulty hitting a target spot marked on the paper. Likewise, in principle, if one could determine all the forces acting upon each raindrop within a rainstorm, the destination of each raindrop could be known. (Assuming that the effect of quantum uncertainty is "normalized.") Thus, our perspective of a rainstorm's randomness is really an artifact of our limited knowledge of the events involved.

Let's go deeper with this principle. If one examines, with a high power lens, a microscope slide of a water drop containing a suspension of particles of appropriate smallness, one can see the jittering motion of Brownian movement as the invisible water molecules jostle the visible particles around. From our perspective, this motion certainly qualifies as random. This "random" process of water molecule movement surely occurs within falling raindrops confined within each drop by the boundary produced by surface tension. And yet the whole raindrop system hits the raindrop's "target." Thus, we see that there can be "purpose" in spite of "randomness" of events. (Interestingly, the Bible claims the natural phenomenon of rainstorms to be God's doing, an outcome of his immanency. See, for instance, Psalm 147:8 and Job 28:26.)

Let us here reflect on the matter of "chaos." We have already considered the basic principle of quantum uncertainty and the decoherence of the involved probability waves. At the superquantum level, certain systems are described as chaotic. Here even very small variants in initial conditions can have major effects on future events. This suggests a process whereby quantum conditions may determine the outcome of superquantum events. And "chaos" itself involves "attractors" and thus the emergence of at least limited order in the system.[17] It would seem that this perception of randomness would also be subject to the principles described above.

We will return to these matters in a different context later, especially the role of "randomness" in genetic mutation.

CHAPTER TWO:
Is Sentience Transcendent?

Is there evidence that a Sentience (cognitive, conscious knowing) transcends Comprehensive Immanent Sentience? Is God truly Infinite?

We humans now recognize that our universe consists of billions of trillions of stars and stretches out for billions of light years. We here postulate that there is an absolute somethingness that is infinite, and that there may be an infinity of universes, that is relative somethingnesses, besides our own. An obvious question that emerges is "If there is a sentience (God) that transcends our universe, and all other relative somethingness, why should God be concerned that we humans, on this tiny planet, be provided evidence of his existence?" Why should it be that we of limited sentience be allowed to conceive of God's infinite sentience? To better understand this dilemma, let us further explore the implications of God as infinitely complex, absolute somethingness.

First, let's consider the following question: Does infinite complexity have boundaries within infinite absolute somethingness? Certainly finite, relative complexities such as ourselves have such boundary. If we postulate "yes," does this not negate the infinity of the complexity (God) and thus place him in the category of relative complexity? He could still perhaps be "a god," but not "the God." So, then, it seems that God must have presence within not only the infinitely complex, but also within the most simple and all in between. Such complexity may well have "degrees," but not boundaries. That is, God must be omnipresent including within the finite relative somethingnesses such as our universe, as we have seen, even at the quantum level. Again, God must be immanent as well as transcendent.

In this context, we see that the significance of our planet is relative not only to the whole universe, but also to the quantum level. We have argued that God must then have a presence, and thus awareness, of events at our level. We have already considered the matter of God's involvement in the outcome of these events. Let us now consider implications of this perspective, and especially how they may relate to transcendency.

Now if God, in his infinite sentience (omniscience) is fully aware of us, but we, in our finite sentience ("miniscience") only speculate as to his existence and nature (as evidenced by the numerous concepts of deity in *Homo sapiens* universal practice of religion), then it follows that God must reveal his transcendent sentience to us at our level if there is to be conscious interaction between us. We obviously are unable to reach his level in spite of our numerous attempts. Has God done this and if so, how? Be aware that here we are dealing with God's infinite, transcendent sentience rather than his immanency so that our science, with its limitation of empiricism, is hardly an adequate means of discovery. But God, in this transcendent sense, could come to our sentience level through revelation. The New Testament of the Bible holds that God has also revealed himself through incarnation in the person of Jesus. Let us first deal with the matter of revelation.

How might revelation take place? Perhaps by direct appearance to humans. But what would infinitely complex somethingness look like? Revelation perhaps by means of God-inspired "visions." Even if this were so, what proof would an individual have that such an encounter was not illusion? We do not here deny that such encounters have occurred, but point out that a more "credible" means of revelation is to be expected.

We humans mutually communicate well, present and future, through use of written language. Our messages are expressions of our sentience, our thoughts on any particular issue. Could God not, through his immanency, "inspire" our thoughts with information that we could not otherwise know? Written expressions of this information would then be a means of communication, even to future generations. As evidence of the accuracy of the revealed information accumulates, the validity of the revelation would be established. What would evidence a God-inspired compilation of such writings? Foretelling events well before their occurrence (prophecy) would constitute such evidence. If, indeed, quantum uncertainty leaves the future open, then accurate prophecy must evidence inspiration of God's intention of outcomes. It would be a revelation of theistic determinism in that, even a complete knowledge of present conditions, could not allow an accurate prediction of future events in an open system independent of God's action.

Another evidence of inspiration recorded in writings would be unwitting allegory. Here an account has an immediate meaning, providing evidence of the time horizon at which it was originated. But also there is an alternate meaning. If this alternate meaning is hidden from the human originators, but later becomes clear as circumstances progress, then, again, it would serve as evidence of divine inspiration. The matter of prophecy, particularly in the Judeo-Christian context, has been dealt with at length elsewhere. (For instance, the New Testament of the Bible explains the fulfilment of Old Testament prophecy. And even more current events, such as the reestablishment of the nation of Israel, can be taken as the fulfilment of prophecy.) Here let us expand our consideration of this form of allegory.

Allegory and Religion

Our quest for examples of divinely inspired allegory needs a context. This context may well involve science, at least in evidencing the validity of hidden meaning in the allegory. However, it must go beyond science in its involvement in that its origination would be transcendent to the nature that science studies.

The obvious context that we seek, then, is religion. Typically, religions involve written compilations that communicate the ideas and information that serve as the basis for a social system. Let us look for a suitable example to analyze.

First, the human religious perspectives that do not recognize the infinite sentience we hold to, but rather are polytheistic, in the sense that no one god is God, would be disqualified as originating from the revelation of true infinite sentience. Of the monotheistic religions, Judaism, Islam, and Christianity, perhaps the latter best conceives of God's immanency. All three recognize God as transcendent. Christianity, then, recognizes transcendent God, the "Father," immanent God, the "Spirit," and a third "person," the "Son," having the function of personal communication with finite, sentient beings and the reconciliation of at least human beings, characterized as flawed. (The existence of other finite sentiences similar to us but from other "worlds" apparently is neither confirmed nor ruled out in biblical writings. However, Ephesians 3:10 speaks of "rulers and authorities in the heavenly realms.") In the Christian perspective, all three "persons" (the terms used refer to human analogy) share sentience, and thus are a single, complete whole. They are a "Holy Trinity," three-in-one. This Christian perspective fits well our concept of transcendent and immanent, infinite, sentient somethingness referred to as God. Therefore, let us use this context to explore for evidence of allegory as a means of communication from God to us humans. (We will leave a similar examination of other religious perspectives to those so motivated.)

Evidence of Divine Communication through Unwitting Allegory in a Biblical Context

The Bible is a compilation of writings originating over a span of several centuries. One expects its accounts to have had significant meaning to each author and his contemporaries. One would hold these meanings as validation of the account's antiquity. However, if these accounts are allegorical, with hidden meanings the significance of which the authors could not have known, but which historical events and progress in humankind's understanding of nature reveal at future time horizons, then they could be considered as evidence of revelation from a transcendent sentience. Let us consider some examples to establish this principle.

First, an example that clearly validates the antiquity of a Bible portion. The Bible is divided into what are termed "books." Let us look at evidence for the antiquity of the Book of Daniel and at an example of allegory found therein. According to the Bible, the setting for the book of Daniel was the ancient city of Babylon, the ruins of which lie near the present city of Hillah, Iraq. At the time of the narrative, Nebuchadnezzar II was king (604-562 BC). Apparently, early in his reign, Nebuchadnezzar had built a magnificent, six hundred room, ziggurat-style palace that dominated the city's landscape. This is shown in a reconstructed cityscape posted as a wall mural at the entrance of the Ishtar gate. (Google: "Babylon, Nebuchadnezzar's palace" to study posted photographs.) This palace was constructed of bricks weighing up to forty pounds. The mortar used was bitumen, an ooze apparently from nearby petroleum deposits. King Nebuchadnezzar apparently had his masons inscribe some of the bricks with messages he, himself composed. These inscriptions are in cuneiform, a writing system developed originally from pictographs, advanced by the Sumerians and eventually adapted as Babylonian. (Encyclopedia International, 1966.)

In 539 BC, Babylon surrendered to Cyrus the Great. Thus, began the demise of the city and, over the intervening centuries, most of its structures have been left in ruins. During this time, nearby residents have plundered the ruins for building materials for their own use.

In the late 1800s and early 1900s, when Iraq (previously included in Mesopotamia), was under British control, British and German archaeologists excavated the Babylon site for significant artefacts. Many of these are now in the British Museum and other museum collections. However, these early excavators overlooked at least one inscribed brick that was incorporated into the wall of an outbuilding in a nearby village. Through a series of events not important to this context, this brick was exported (about 1935) with a proper permit, signed by the then director of antiquities in Iraq, Mr. J. Gordone. It is presently in a temporary location, eventually to be placed in a museum in the United States. I have had occasion to examine this brick at length. (See figure one.)

The cuneiform inscription of this brick has been translated, first apparently by a German archaeologist before it was exported, and again recently by the Turkish archaeologist, Veysel Donbaz. These translations agree well. Following is the translation in part.

"Nebuchadnezzar, king of Babylon, son of Nabopolassar, King of Babylon, am I. I built a palace for my royal dwelling in the district Kadimgira which is within Babylon. I raised it mountain high with bitumen and kiln-fired bricks. At your command, oh Marduk, wisest of the gods, the house which I built, may I prosper its abundance may I reach into old age in it."

This hand-inscribed message is taken as evidence that this was a dedicatory brick for Nebuchadnezzar's palace and that the message was originated by Nebuchadnezzar himself. Other similarly inscribed bricks and brick fragments are in the museum collections mentioned above.

Now let us turn to the book of Daniel in the Bible. In chapter four, Nebuchadnezzar makes a statement in the first person: "To the people, nations and men of every language, who live in all the world," verse 1. Then in verse 4: "I, Nebuchadnezzar, was at home in my palace, contented and prosperous . . ." (NIV Zondervan. Some translators use "flourishing.") He then goes on to document one of his visions. According to F. F. Bruce[18], this section of chapter four in our earliest manuscripts is in Hebrew. Presumably it was first written in Aramaic, the common language of the Babylonians. In comparing the brick inscription with this statement ("May I prosper its abundance" compared to "Was at home in my palace, contented and prosperous . . ."), we have good evidence of identity of authorship. This also constitutes evidence that the translations are accurate.

The book of Daniel contains several prophecies involving allegory. One in particular is of note here. In Daniel, chapter 2:34-35, as part of a vision Nebuchadnezzar had, a rock is described as crushing a statue, the parts of which represent various "kingdoms." In Nebuchadnezzar's vision, the rock remains, replacing the destroyed kingdoms. Daniel interprets this to signify a permanent "kingdom" set up by "the God" of heaven (verse 44). F. F. Bruce, ibid., 856, points out the revealing of the hidden meaning of this allegory in a statement by the apostle Paul (I Corinthians 10:4): "the Rock was Christ" whose advent introduced a new (spiritual) kingdom truly universal and enduring.

In this example, we have seen evidence validating the antiquity and authorship of a part of biblical scripture. In this same biblical document, we have seen an allegory with a proximate, mystical meaning, and a hidden, ultimate meaning clarified historically. Undoubtedly, there are numerous other archaeological validations of other portions of the Bible. But let us now turn to biblical statements pertinent to current science. What allegories are to be found here?

Science and Biblical Allegory

The opening accounts in the first book of the Bible, Genesis, certainly give convincing evidence of their antiquity. For instance, the creation account itself has a somewhat parallel account in what is taken as contemporary literature, being inscribed in cuneiform on seven clay tablets which were found in the twelfth century BC ruins at Nineveh. This, the Enuma Elish Mesopotamian version, is modified from earlier Sumerian versions. Its god hero is Marduk, the prominent god of Babylon (named on the brick inscription cited earlier). However, this parallel statement differs significantly from the Genesis account in that several gods are involved rather than the God. Although its format is sufficiently similar to that of the biblical creation account to establish the two as contemporary, there are many stark differences. For instance, the Enuma Elish account is primarily of "family feuding" among the gods. Eventually, Marduk dominates. In doing so, he defeats and kills Tiamat, goddess of the sea, chaos and threat. He splits her body "like a shellfish" into two parts. Half of her he sets up as a "covering for heaven." He "pulled down the bar and posted guards." He "bade them not to allow her waters to escape." This all from tablet four. In tablet five, it says "taking the spittle of Tiamet, Marduk created . . . He formed the clouds and filled them with water, the raising of winds, the bringing of rain and cold . . .." Note that although "waters" are kept in the "heaven covering," rain is understood to come from clouds.[19]

In the biblical account, the "days" of Genesis one would surely have been thought of as actual days in the contemporary context. However, if allegorical, deeper meaning is expected.

It is interesting to note that, if taken literally, the incidents in Genesis where God questions Adam and Eve of their whereabouts (3:9) and where God asks Cain what he had done to Abel, his brother (4:10), characterize God as less than omniscient. However, in Psalm 139, we see God as not only omniscient, but also omnipresent. Thus, the Genesis statements should be regarded as allegorical rather than literal.

Suppose that the creation account in Genesis were scientifically complete by modern standards, for instance, acknowledging the vastness of space and details of galactic events. Would not this be taken as evidence of an ancient, advanced science? Rather, an ancient account taken as unwitting allegory would have a proximate, apparent meaning which validates its antiquity. However, its hidden, ultimate meaning, perhaps now revealed by our advanced science, would validate the original statement as revelation because the ancients who wrote it could not have known this meaning. Thus, the days of Genesis, chapter one would have double meaning if taken as allegorical.

Again, what is meant in Genesis when God "speaks" his creation into existence (chapter one)? Could this, too, be an incidence of allegory? In the account, there is no air yet to carry sound waves. Nor is there anyone to talk to. But could not another meaning be that this Genesis statement refers to God actualizing his intentions? Recognize that communicating one's intentions (with the expectation that they will be actualized) is a purpose of speech as we know it. So the statement that God "spoke" is fitting enough.

Note, however, that the text does not otherwise specify the means whereby God creates, that is, actualizes his intentions. Obviously, many have understood the meaning of the Genesis statement to be that God opened his "mouth" and uttered words when he spoke.

But perhaps this is only anthropomorphic in concept. Remember that herein is the function of allegory. But then, does our science clarify a hidden meaning here?

We saw earlier how Psalm 139 clarifies God's nature. I suggest that the means of God's creative action is clarified in Psalm 104:27-30. This passage can be taken as a beautiful, poetic expression of immanent God at work in nature—work that he continues as he sustains all of nature. (Note that the passage does consider God's creative acts as ongoing. See also John 5:17.) Later we shall reflect on what science has revealed as to this process.

In summary, the initial passages in Genesis not only provide evidence of their antiquity, but also, among other matters not considered here, establish a fundamental concept of God and his involvement in our universe—this concept being of the God, as opposed to the then contemporary, polytheistic perspective. We see that the God is transcendent to our universe (e.g., he exists "before" it does), and he is immanent in the universe (e.g., Genesis 1:2 "the Spirit of God was hovering over the waters"). He is one, yet with three "persons," the third person's interaction with human kind foretold as prophecy (Genesis 3:15). An offspring of Eve would, in time, "crush the serpent's head," a promise fulfilled in Christ's victory over Satan (Romans 16:20). Thus, the Bible's monotheistic concept of God fits the requirement of infinitely complex somethingness, transcendent as well as immanent. It is also clear from the Genesis account that the laws governing nature are, indeed, God's laws, being an essential aspect of his creation.[20] Further, we read in Revelation 4:11 the following:

> "You are worthy, our Lord and God
> to receive glory and honor and power,
> for you created all things,
> and by your *will,* they were created
> and have their being."

More Examples of Biblical Allegory

Here we will look for evidence of transcendent God as the source of revelation in these examples of biblical allegory. Our considerations may be lengthy, involving both physical aspects of nature and, especially, biological aspects. First, let us consider physical examples.

CHAPTER THREE:
Biblical Allegory in Physical Science

Physical events within our universe have been postulated to involve, at a fundamental level, such basic units of matter (matter particles) as the quarks and leptons (including neutrinos, electrons, muons and taus) that make up atoms. Four basic forces are known that act upon ordinary matter, the strong and weak forces, electromagnetism, and gravity. These, in turn, are known or postulated to be mediated by force field "particles," the gluon, boson, photon, graviton, and Higgs particles. Apparently, each of these entities, according to quantum mechanics, may also have a wave function. The manner of interactions involved among these physical entities are known as the laws of physics (fide the standard model).

An alternative perspective known as string theory holds that the basic components of ordinary matter are multidimensional, vibrating "energy" strings or membranes rather than actual particles. These matters are expounded in various physics texts which can be consulted for further development of this context. (See also Lisa Randall, *Knocking on Heaven's Door* [New York: Harper Collins, 2011], fig. 46, 254.)

The "Big Bang" cosmological model holds that these laws and forces, as well as the one time and three spatial dimensions characteristic of our universe, were the outcome of the fate of the singularity involved in the origin of our universe. Presumably, other universes may be characterized by different laws and forces if not dimensions. Steinhardt and Turok's ekpyrotic (or "big splat") model, an alternative to the inflationary (or "big bang") model, allows stability of these laws, forces, and dimensions through perhaps an infinity of cycles.[21] Later we will consider the pertinence of this model to the biblical perspective.

Certainly, the physics of our universe gives evidence of order emergent from the disorder of quantum uncertainty. But is this order evidence of sentience? Again, one might assume that it "just is," without cause, and take by faith that absolute somethingness, transcendent to our universe, and with potential for infinite sentience, "just is not."[22] The consequences of these conflicting perspectives are basic, indeed. On the one hand, absolutely everything matters and has purpose. On the other hand, nothing matters in an absolute sense. From the latter perspective, our existence certainly represents the winning of a "cosmic jackpot" (See Paul Davies, *Cosmic Jackpot* [Boston: Houghton Mifflin, 2007]). From the former, such orderliness represents God's actualization of his intentions, actualization of the "eternal potential" characteristic of theistic determinism. And if this is so, accurate revelation in the sense we have described must, indeed, evidence origination from sentience transcendent to our universe with its open future, that is, from the divine designer himself.[23]

Thus, the question we deal with here is "can the physical aspects, including the dynamics, of what we observe in our universe clarify hidden meaning in biblical statements thus establishing them as unwitting allegory and thereby evidence this divine revelation?"

Our search for biblical allegories requires sorting through an abundance of examples that evidence the antiquity of the various accounts. For instance, Psalm 104:5 states: "He set the Earth on its foundations, it can never be moved." Perhaps this could have the hidden meaning of the planet Earth's stable orbit. However, the Earth certainly moves through space. In Job 38:13, we read that the Earth has "edges." In Revelation 7:1, angels are stationed at the "four corners" of the Earth.

One wonders how this description could ever "square" with the reality of the planet Earth's spherical shape that we now understand!

The ancients apparently had the conception of the Earth as flat and the sky, or heavens, above the Earth much like an inverted bowl. In this conception, there was the "true" heaven somehow above the bowl of the sky. Apparently, access required some sort of penetration into the true heaven. See fig. 2. And yet our telescopes provide quite a different perspective.

In the Bible, heaven and Earth are often contrasted. 1 Kings 8:27 states "But will God really dwell on Earth? The heavens, even the highest heaven, cannot contain you." (The writer is making the point of God's "infinity" not to be limited to" relative somethingness.") But what is meant by "highest heaven" as opposed to "heaven" here? How might this ancient perception be allegorical? What hidden meaning can there be?

As noted earlier, Steinhardt and Turok, in their book, *Endless Universe*, published in 2007, expound on their ekpyrotic model of the "universe." This model holds that our universe is matched to another universe, apparently also with three spatial dimensions, but separated from ours by a fourth spatial dimension. Both universes are three-dimensionally flat. Their attempt to illustrate this relationship is most interesting. For instance, on pages 158 and 159, they provide an illustration showing the cyclic nature of their model. The two parallel universes are represented (allegorically) as two flat sheets each with four corners and with edges. Understanding that their model involves a three-dimensional, parallel relationship, it is clear that the figures only represent a reality difficult for us humans to conceptualize. We would hardly condemn them for inaccuracy! Perhaps the biblical account given by the ancients is similarly intended, in its hidden, divinely inspired, allegorical meaning, to represent this relationship.

From the human perspective, our planet, Earth, is part of our extended universe. Thus, perhaps one meaning of "Earth" in the biblical account may refer to our whole "flat" universe. (Apparently, the Greek word, "kosmos," used in some of the biblical passages can mean either the planet Earth or the whole universe.[24] In the ekpyrotic model, passing from Earth (our universe) to heaven (the other universe) would involve passing through the fourth spatial dimension that separates the two universes (rather than travelling faster than the speed of light to somewhere beyond the range of our Hubble telescope).

And what is the nature of this fourth dimensional barrier? Perhaps Denis Lamaureux comes to his conclusions prematurely when he holds that there is contradiction between the Bible and science in regards to the origin and structure of the heavens, there being no scientific evidence of a solid, heavenly structure upholding water. (*Perspectives on Science and Christian Faith* vol. 60, no. 1, [March 2008]: 10.) He points out that the Hebrew word *raqia* does not refer to the troposphere or outer space as many have interpreted it. Rather this Hebrew word refers to something "flattened out and solid" rather than a broad open space such as the atmosphere (ibid., 5). In the context suggested at this writing, we see amazing accuracy in an unwitting allegory. What could be more "flattened" than a fourth spatial dimension only ten to the minus 30 centimeters "thick"?! (Steinhardt and Turok, *Endless Universe* pg 139.) "Solidity" implies impenetrability. What is more impenetrable from the natural human perspective than spatial dimensions other than our three? And that this "firmament" separates "waters" is a most astounding, unwitting allegory as we shall explain shortly. In light of these insights, perhaps biblical and scientific accounts are, indeed, amazingly concordant!

And what of the Bible's prediction of a "new heaven and new Earth" (e.g., see Isaiah 65:17 and 66:22 and Revelation 21:1)? Interestingly, the ekpyrotic model predicts this as well. At intervals of a trillion years or so, the two universes, or in string theory terminology, "branes," collide with each other, rebounding with renewed energy.

As mentioned earlier, the energy available for this is apparently infinite in the form of negative gravitational potential energy so that, with each cycle, each universe is fully rejuvenated. (See Steinhardt and Turok, ibid., 191-192.) Entropy occurs, of course, but is kept "diluted" by the continued expansion of each flat universe (brane), the expansion being driven by dark energy which is derived from ordinary matter and energy, this occurring in the intervals between the brane collisions.

To be sure, there are phases of this model that would seem quite inhospitable to life forms. Perhaps there are other universe systems with alternate phases that can somehow be accessed. Or perhaps the Bible's "New Jerusalem" emanating from "heaven" and distinct from the "Earth," is to be a haven (a cosmic "Noah's ark") during these rejuvenation stages? (See Revelation 3:12 and 21:2.) In John 17:3, Jesus says in a prayer to God, the father, "now this is eternal life: that they may know you, the only true God . . ." Again, in Hebrews 1:11, we read: "they [heaven and Earth] will perish, but you remain . . ." In verse 1:12,: "like a garment they will be changed. But you remain the same, and your years will never end." Those, then, who "know God," who is eternal, likewise have eternal life, which would necessitate survival of the collision phase of each cycle.

Apparently, the collision phase of a cycle results in the generation of "a searing white light" (Steinhardt and Turok, ibid., 193). The separation distance in the fourth dimension between the two universes is, again, about 10 to the minus 30 cm. (ibid., 139). Yet the energetics of the interaction, when the spring-like force connecting the two universe branes through the fourth dimension causes their encounter (collision), is adequate to create matter and radiation with "enough excess kinetic energy left for the branes to bounce back to their original positions . . ." (ibid., 191). Apparently, the interaction of the components of the separate branes upon contact (remember that this is a three-dimensional interaction as the fourth spatial dimension contracts to zero), generates these energetics, including the emission of the searing white light. Would this ekpyrotic model in some way allow premature interaction of the components of the two universes on a limited scale? If so, would this interaction also result in the emission of brilliant light? Thus, the biblical accounts of what might be taken as such encounters, such as the appearance of angels to the shepherds at Christ's birth, should be characterized by unusually bright light. Luke 2:9 states "an angel of the Lord appeared to them, and the glory of the Lord shone around them . . ." (New International Version, [Grand Rapids: Zondervan, 1985]). In Matthew 28:2-3, an angel's appearance is described as "like lightning" In 2 Kings, 6:17, we read of "chariots of fire." Various other biblical accounts do not attribute unusual light emission associated with the appearance of angels, especially in the Old Testament. Indeed, if this phenomenon of light emission is due to passage through the fourth dimension, it may be limited to the actual period of passage. Also, the light may be emitted only on the side of the fourth dimension receiving the transmitted entity. Assuming this model to be actual, both ekpyrotic and biblical versions, the biblical account of passage from our universe into the other has the entities fading away as into a "cloud" as seen by an observer on our side. For instance, at the "transfiguration" (Mark 9:3-5) "Elijah and Moses" appear associated with "dazzling white" (of at least Jesus's garments, perhaps a reflection), but they disappear into a cloud that "enveloped them." Perhaps the cloud effect was rather due to the fading of their image as they passed back through the fourth spatial dimension into what we will think of as the "heaven" brane.

In the ekpyrotic model, the only entity that normally can pass from one brane to the other is the graviton. Apparently, the graviton is unique in being a "string ring" with no open ends attached to the brane's dimensions so that it is free to pass between branes through the fourth dimension. (See Steinhardt and Turok, ibid., 125, 139.) For instance, dark matter is apparently detectable by us only by its gravitational force, leading to speculation that it may actually be the matter within the universe alternate to ours.

If, indeed, biblical accounts of passage of entities (usually "beings") between these branes are true, it would not be surprising if the point of passage were not necessarily at our ground level. Such passage points might well be "in the air" where they presumably would be supported by gravitational force from the other brane through the fourth spatial dimension. Obviously, such entities would have a different nature than those in our universe so that, perhaps, gravitational force from the alternate universe would dominate over our gravitational force in these encounters. Yet, gravitons are apparently the same in both branes. (Such is the "adventure" in acts of reasoned speculation!)

Stephen Hawking, on pages 180 and 181 in his book *A Brief History of Time* (10th anniversary edition, [New York: Bantam Books,1988]) points out problems with universes having more than three spatial dimensions. For instance, he states "the electrons (of atoms) would either escape from the atom altogether or would spiral into the nucleus." Thus, "one could not have atoms as we know them." However, might it be that atoms of both of the ekpyrotic three-dimensional realms could coalesce into an entity transmissible between the realms? In our experience, when a three-dimensional object passes through a two-dimensional plane, both share a two-dimensional space. Thus, if two three-dimensional objects coalesce, might they not share the same space, thus bringing both realms into register? Perhaps then the object would have the capabilities of a six-dimensional entity, but yet retain the integrity of its atoms. Or, perhaps, the combination of the two three-dimensional components of atoms (quarks, etc.) take the form of graviton-like rings so that their otherwise free ends attach to their "partner" components from the other realm and thus, free them from attachment to the spatial dimensions of either brane realm. It seems, though, that this would compromise their atomic characteristics in other ways as well.

Hugh Ross, in his book *The Creator and the Cosmos*, (3rd ed. [Colorado Springs: Nav Press, 2001] 111), gives an explanation of the biblical account of Christ, in his resurrection body, appearing within a closed room. We easily visualize accessing a two-dimensional space enclosed by a barrier such as a line by simply passing around the barrier line via a third-dimension. Likewise, access to an enclosed three-dimensional space could be made by passing "around" the barrier, e.g., an enclosing wall, via a minimum of six spatial dimensions, the precise number resultant from a "coalescence" of the dimensions from each brane in the ekpyrotic model!

In considering these matters, one remembers that the ekpyrotic cosmologic model, at this writing in 2008/2011, is still largely hypothetical and the explanation of the biblical accounts we give is speculative. However, the similarities are certainly suggestive. One is almost tempted to make a forecast ("prophecy").

But there is more! (Warning! You may find the following insights almost "hair raising.") We have already mentioned the biblical prediction of a "new heaven and a new Earth," suggesting that the use of the term "Earth" in this context could be allegorical with a second meaning referable to our brane of the ekpyrotic model, with heaven perhaps referable to the other brane.

And we have recognized that the ekpyrotic model predicts the conversion of matter to dark energy, driving the expansion of the branes that eventually become "empty, flat, and parallel" (Steinhardt and Turok, ibid., 157). This condition precedes the collision (or "big splat") phase of the trillion-year cycle. The collision starts the renewing phase eventually resulting in both branes again dominated by matter which forms galaxies and stars. If this model is correct, we see a rather amazing match to a biblical statement, Hebrews 1:11, "they [heaven and Earth] will perish . . ., they will wear out like a garment." Verse 12 states, "like a garment they will be changed." One wonders why the ancients would think that especially the heavens should need to be changed or renewed. Again we see in Revelation 21:1, "then I saw a new heaven and a new Earth, for the first heaven and the first Earth had passed away . . ." The conclusion that this is an unwitting allegory validating the divine inspiration of these scriptures is hard to avoid!

Note that Hebrews 1:12 also states "you will roll them [heaven and Earth] up like a robe . . ." According to string theory, at least some dimensions of this universe are "rolled up" into what are termed Calabi-Yau entities. Further, Job 9:8 reads "He alone stretches out the heavens . . ." It is now well-established that at least our universe is expanding. The ekpyrotic model holds that both branes of the system continue to expand. And in Job 10:22 we read, "to the land of deepest night, of deep shadow and disorder, where *even the light is like darkness*." Again, in the ekpyrotic model, there is a phase in which "matter and radiation (light) are diluted away" (Steinhardt and Turok, p. 157) being converted to dark energy which "becomes dominant" (ibid. p. 62). How could such cosmic principles be so clearly stated in the context of an ancient society, who could not have known their allegorical meaning, unless by inspiration from a transcendent sentience! We shall return to more implications of these observations shortly.

There are undoubtedly other biblical examples of allegory, such as the statement in Job 26:7 describing the Earth as "hung on nothing." But, perhaps those we have chosen will suffice to establish the success of our quest for inspired biblical allegory in the physical realm. However, there is one more which we cannot omit. Here especially comes the astounding part.

There has been considerable disagreement among biblical scholars as to the intended meaning of the creation account in the first chapter of Genesis. There is an abundance of literature expounding upon this disagreement, so we shall avoid the details in this context. Until the development of the already referenced Steinhardt and Turok's ekpyrotic model and the publishing of their book Endless Universe in 2007, the following observations on the allegorical significance of Genesis, chapter one would not have been evident.

CHAPTER FOUR:
The First Chapter of Genesis and
the Ekpyrotic Model

Let us now progress through the creation account of the first chapter of Genesis, stage by stage. These stages are termed "days" in the account and must certainly have meant ordinary days to the ancients who recorded the statement. But what of hidden allegorical meaning? Will the stages of the cyclic ekpyrotic model give us any insights?

"In the beginning God created the heavens and the Earth." So states Genesis 1:1, the first statement of the Bible. This obviously is not the beginning of absolute somethingness which, as we concluded earlier, is eternal as God is eternal. Rather, it would refer to the relative somethingnesses that make up our present universe and "heaven." The cyclic model suggests that our turn in the cycle, and in a sense, our relative somethingness, begins with a collision of the branes, the "big splat," or more phonetically fitting, the "flat splat." As mentioned earlier, the ekpyrotic model holds that a "searing white light" is produced at this event. But apparently the photons composing this light interact with free electrons in the resulting plasma and are "scattered" within the plasma itself, there being no empty space in which to radiate.

"Now the Earth was formless and empty, darkness was over the surface of the deep . . ." we read in Genesis 1:2. Certainly the "hot (dark) plasma" of the next stage of the ekpyrotic cycle is formless and empty of matter. Continuing the quote: "and the Spirit of God was hovering over the waters." Might the "waters" refer to the infinite, negative gravitational, potential energy of the ekpyrotic model? Perhaps, but more likely "waters" refers to the plasma of this stage of the sequence. The ancients would have no separate word for plasma which would be a concept unknown to them.[25]

Genesis 1:3 states that God said, "let there be light!" The question has often been asked, why light, before the existence of the sun and other stars? Now the answer is apparent! The next stage of the cyclic model is characterized by the branes' rapid stretching that allows the plasma to cool. And when the plasma cooled sufficiently for its free electrons to bond to atomic nuclei, thus producing complete atoms, the plasma became transparent. At that stage "all the light energy generated in the beginning was allowed free" (Clegg, *Before the Big Bang* [New York: Saint Martin's Press, 2009] 108, but referring to the big bang model). In Genesis 1:4, "light being separated from darkness" seems beautifully concordant with this phase of the ekpyrotic model. Note that the "day and night" of verse five precedes creation of the sun.

It is noteworthy that Steinhardt and Turok indirectly recognize this concordance when, on page 193 of *Endless Universe* (New York: Doubleday, 2007), they quote from Isaac Asimov's, *The Last Question*, (New York: Columbia Publications, 1956). Here the character AC says, "let there be light!" and there was light, this the familiar phrase from Genesis 1:3

Genesis 1:6-7 reads: "And God said, Let there be a firmament between the waters to separate water from water. So God made the firmament and separated the water under the firmament from the water above it." The ancients apparently concluded that the blue of the sky indicated the presence of water in that it matches the blue of bodies of water on the ground. After all, water does fall from the sky on occasion. Of course, it is well understood now that the blue of the sky, indeed, is not water, and that rain originates in clouds. Thus, this biblical statement has been difficult to explain.[26]

However, again, if one takes the term "waters" to refer to the plasmas of the two branes which are separated in the rebounding phase of the ekpyrotic cycle, this statement makes a sense that the ancients could not have known! Note that the Bible states that the firmament separates "water from water" and that the cyclic model holds that the branes, still loaded with plasma, are separated by the barrier of the fourth spatial dimension.(Note also Psalms 148:4 which implies "waters" in the "highest heavens.")

In Genesis 1:9, the "water under the sky" (or firmament) is "gathered to one place" whereupon "dry ground" appears. In the ekpyrotic model, plasma and the resultant radiation "convert" to matter which clusters to form galaxies and stars, perhaps allegorically represented in the biblical statement by the "dry ground." At least during a part of this phase of the cyclic model, the planet Earth exists and is able to support vegetation. However, it does appear that the ancients got the creation of plant life out of order, coming before the creation of celestial bodies as described in Genesis 1:14-17.[27] All would occur during the same phase of the cyclic model, however, as would the appearance of animal life described in the remaining part of the creation account.

We see, then, that the stages of the ekpyrotic model amazingly fit the first six days of the Genesis account. In day seven, God rests from his creative action. Some take this to mean that God has ceased involvement in the matters of our universe. However, in Colossians 1:17, we read "in him all things hold together." This certainly implies that God continues his involvement in nature. In John 5:17, Jesus says "My father is always at his work to this very day . . ." Perhaps then, God's "day of rest" signifies that his creation of entity categories is completed and therefore, he refrains from the "work" of adding more. Chapter two of Genesis gives another version of creation which we will return to later.

As we have seen, the ekpyrotic model predicts a new cycle in the future in which both branes or universes ("earth and heaven") will be made new, also in keeping with the biblical model. The events described here are diagrammed in figure 3 in an attempt to clarify our perspective.

Our considerations of biblical allegory from a physical perspective are now complete. Hopefully, they are sufficient to demonstrate the richness of biblical allegory from a physical science perspective. Let us now look for such allegory from a biological science perspective.

CHAPTER FIVE:
Biblical Allegory in Biological Science Context

Part I: Population Dynamics

Psalm 104:25, "There is the sea, vast and spacious, teeming with creatures beyond number, living things both large and small." 104: 27-30: "These all look to you to give them their food at the proper time. When you give it to them, they gather it up; when you open your hand, they are satisfied with good things. When you hide your face, they are terrified; when you take away their breath, they die and return to the dust. When you send your Spirit, they are created, and you renew the face of the Earth."

Mathew 6:11, "Give us today our daily bread."

Mathew 10:29, "Are not two sparrows sold for a penny? Yet not one of them will fall to the ground apart from the will of your Father."

Luke 12:27, "Consider how the lilies grow . . . not even Solomon in all his splendor was dressed like one of these. If that is how God clothes the grass of the field . . ."

From these statements, it is evident that from a biblical perspective God does have an ongoing, active involvement in the events in nature—especially the passage from the Psalms portrays God's active involvement in the dynamics of ecology. Newly created forms are the result of selective survival and are at the expense of extinctions. And all of this is with God's awareness and control. Biologists speak of the selective force as natural. Psalm 104 declares it as the action of God's spirit, that is, his immanent presence in the natural events here on our planet. Note that these events must needs include not only the genetic mutations providing phenotypic traits and reproduction modes that determine their expression, but also the physical parameters that constitute the selective mechanism. All of this may be considered natural, but there is no biblical distinction made between natural events and God at work. It is quite biblical to consider God's involvement in nature, natural.

Thus, the dynamics of living systems are certainly a part of the order that constitutes God's "footprint," evidencing his existence. And the science of biology is part of the documentation of this order and hence a significant contribution to understanding this evidence. The biblical statements given above take on meaning not apparent at their writing and thus, again, function as un-witting allegory. And from this perspective, there is no problem in recognizing the dynamics of change in living systems as God ordained. The process of chronological modification (evolution) is, indeed, a true witness to God's act of actualizing his intentions. Later, we will consider how the implications for human origin from this perspective fit the biblical context.

The Intelligent Design of Chronological Modification

Let us now consider the basic mechanism of organic evolutionary change as it is now understood. First, we shall outline the principles involved and how they may relate to a biblical context, then later deal with examples in some detail as they illustrate the principles. Most of the aspects we will encounter are well-covered, in different context, in various text books dealing with the corresponding subjects, such as molecular genetics, genomics, population genetics, animal and plant reproduction, paleontology, and other aspects of the science of evolution. (Note that we omit the philosophical perspective of "evolutionism" which is certainly not a part of the science of evolution.) Our present intent is not to comprehensively recount the details of these subjects, but rather to draw on them as they fit our context.

Earlier we considered the significance of the biblical perspective of God's involvement in natural events such as rainstorms. The primary significance of this example, at least in our present context, would be God's control of the fate of each raindrop, so that from his omniscient perspective, superquantum chance is not reality. God's perspective is in contrast to our finite perception. The rainstorm waters our garden, but also has many other functions, all of which are objectives from God's comprehensive perspective.

Now suppose that we liken changes in the gene pool of a population of organisms to the random pattern of raindrops. That is, the pattern resulting from the combination of all the gene mutations being "random" like the impact pattern of raindrops. We could also liken the parameters of the genetic "needs" of a population of organisms to the boundaries of our garden plot (see figure 4). It is evident that the positions of the boundaries determine the usefulness of any given mutation within the gene pool.[28] But does that mean that mutations falling outside the immediate-need boundaries are waste? Note here that usefulness is thus relative to the boundaries. If the boundaries change, different mutations become useful. Also, it is pertinent here that sexual reproduction and the dominance vs. recessive relationship of gene alleles provide a system whereby at least mutations resulting in recessive gene alleles falling outside of the immediate-need boundaries can be accumulated as a reserve within the gene pool and made available as needed within altered-needs boundaries.[29] In large populations, outbreeding dominates so that recessive alleles only infrequently show in phenotypes.[30] But if they do show and give survival disadvantage, they tend to be eliminated. Also it is expected that some DNA alterations, such as the deletion of essential genetic information, would result in malfunctions and elimination of individuals. All of this may function as part of the negative feedback keeping a population from exceeding its carrying capacity and therefore is not "bad" in the eventual outcome. In nature, "defective" individuals may well constitute the food supply of predators. Remember that God feeds even hungry lions (Psalm 104:21).[31]

Now when directional selection occurs on a population as its environment changes, requiring adaptation to an altered ecological niche (the "genetic-needs boundaries" shift giving advantage to different genetically determined traits, see figure 4), the importance of sexual reproduction is evident. Here there is constant shuffling of genes, allowing allele combinations which now may have a selective advantage whereas they were at disadvantage in the old context, that is they had previously fallen outside, but now fall within, the "needs boundaries." (Using our original example, rain that had fallen on the walkway now falls within the new garden boundary and thus is now considered useful from our perspective.) In a large population where inbreeding is minimal, mostly dominant alleles would be affected, the recessives seldom being expressed in phenotypes. However, there are circumstances in which the recessive reserve of alleles can be accessed as we shall explain shortly.

First, let us consider the dynamics of this process of a population's shift in adaptive phenotype. Note that alleles that were of low adaptive value may now fall at or near the new adaptive optimum or peak. Differential survival of individuals making up the breeding population would, relatively quickly, result in enhancing the frequency of these phenotypes (and their gene base) in the population. However, a further shift in the range of phenotypes is now dependent upon new mutations producing new alleles in the gene pool. (These mutations may have occurred, on occasion, prior to the shift in niche, but, at least if they are dominant alleles, would have been eliminated by the selective process then at work.) Consider a generalized example. Wet, adapted individuals would do poorly in a dry climate, but have an advantage if the climate were to shift to wet. Hence, one expects concomitant shifts in the populations phenotype frequencies as the climate changes.

Some people, usually non-biologists, have interpreted the pause in phenotype shifting (due to the "need" for new mutations contrasted to the reshuffling of alleles already present in the gene pool, the process known as "microevolution") as somehow being a permanent boundary beyond which a species cannot change. We see here the conceptual fallacy of this perception. Later we will consider pertinent examples. (Also see appendix I.)

Also, it is of interest in this context that even mutation rates apparently have an adaptive constraint. If they are "excessive," a population's fitness norm may break down. Agriculturists have successfully increased mutation rates, for instance, by using radiation, but do so on relatively small samples of individuals which are isolated from the parent interbreeding population. Most of the resulting mutations in the treatment cause excessive damage, but the perpetrators are able to select out the occasional mutation giving the desired trait. These individuals are then subjected to a breeding program establishing the desired cultivar. Apparently, mint varieties with high oil content have been produced in this way, as an example. However, we would expect the process to be much slower in nature due to the slower mutation rate which allows the "defective" alterations to be selected out without overwhelming the gene pool and yet preserving the occasional useful mutation. (Remember that, from a non-finite perspective in natural systems, "defective" individuals may still have a purpose, perhaps, as we suggested earlier, at least for nonhumans, as food for predators.)

Earlier we suggested that there are other circumstances in which a population's recessive allele gene pool reserve may be accessed. Any situation in which inbreeding rates are increased would tend to accomplish this. For instance, such a situation would occur in a breeding population that has been brought to a size "bottleneck." The increased inbreeding favors homozygosity of recessive alleles so that their traits are expressed. If these traits give good advantage, they may well become "fixed" in the gene pool. (That is, their dominant allele counterparts may be lost.) Note that this may occur at whichever gene loci happen to have recessive alleles in the surviving parents.

Of course, a population size bottleneck means that the number of individuals, each with a maximum of two alleles for any gene (because of diploidy), is reduced and hence much of the recessive allele reserve pool may be lost, especially if the size reduction is "excessive." Therefore, it is apparent that there must be a population size threshold at which inbreeding reaches a significant frequency and yet an "adequate" reserve of recessive alleles remains.

Another circumstance under which the recessive allele pool reserve can be accessed is where wandering individuals, such as fertile females, establish satellite populations. Again, inbreeding of offspring would be at a high frequency. And if there continues to be occasional migrants from the parent population, male or female, there would be continued input from the parent pool of recessive alleles. Over time, this situation could give access to the whole recessive gene pool. The results in the satellite population may be a significant shift in trait frequencies, including the frequent expression of the enhancing recessives. Especially this expression of recessive traits may result in the new population "jumping" to a different adaptive peak. If isolation of the satellite population persists and even increases over time, and selection pressure favors a different "direction" than in the parent population allowing the accumulation of new mutations, then the satellite population may come to occupy a new ecological niche.

Again, let us use our example of wet versus dry adaptation. Say the large parent population is dry-environment adapted. The frequency of individuals with the needed traits constitutes the peak of the population curve. Offshore is an island with a very wet climate. A wandering, perhaps wind-blown, fertile female colonizes this island. Her offspring inbreed, allowing recessive traits to show in their offspring. If these traits enhance the satellite population's adaptiveness to wetness, its adaptive peak will have shifted compared to that of the dry adapted parent population. Furthermore, any new mutations that enhance adaptiveness to wetness would now have high selective value, even though in dry country they would be outside the "genetic needs" boundary. Thus, the needs-boundaries are different on the island.

Suppose now that the island's climate changes to dry, that is, the "genetic needs" boundary shifts again. Obviously, selection pressure now changes direction, favoring traits that may still remain in the gene pool which enhance dry adaptation. However, to the degree that wet-adaptive recessive traits have become fixed, there will be limits to a population's "immediate" response to this shift in directional selection. Even so, one expects that there will be a shift in frequency of traits back toward that of the mainland, dry adapted population. (As an example, apparently finch beak sizes in the Galapagos Islands fit this adaptive dynamic.)[32]

Part II: Scientific Concepts of "Species"

We have reviewed the basic principles involved in adaptive dynamics as they are presently understood. We have dealt with the principle of what has been termed "microevolution" and with a situation that could be classed as "macroevolution," e.g., the jump in the adaptive peak in a newly founded satellite population. But what of "permanent" divergence of populations in what is known as "speciation"? Here we must first set a context. What is meant by the term, "species"?

In view of the dynamic characteristics of environments on planet Earth, perhaps it is not surprising that successful living systems have concomitant adaptive dynamic. We have traced some basic principles of this process above. Perhaps from this it is evident that a given interbreeding population, at a given time horizon, could be at any of the stages considered. And yet, we humans like to neatly "pigeon hole" our concepts. Hence, it is not surprising that we impose concept categories on what we see in nature. The concept of "species" is no exception. As we attempt to apply this category to the reality we observe in nature, we find complexities that make the effort difficult. Biologists hence have several applications of the term.

In a functional sense, the "biospecies" concept is basic. This concept holds that an interbreeding population, as distinct from other interbreeding populations, is indeed a distinct species. We will elaborate shortly. But first, let us acknowledge situations where such a distinction is not applicable or is difficult to determine.

Suppose we are dealing with populations that reproduce without breeding, a situation known as "parthenogenesis." For instance, in my specialty genus *Helophorus* (family Hydrophilidae), there is a "species," *H. orientallis*, which, at least in North America, rarely produces males. Here unmated females are quite fertile. There are obvious advantages such as the conservation of energy otherwise spent in courtship and copulation and the conservation of resources that males otherwise would consume. (It seems that this would be "the way to go" if indeed, species have only degraded since an original perfect condition at non-natural creation.) But in the context of dynamic ecology, there are disadvantages as well. To the extent that interbreeding is lacking, the process of "genetic shuffling" that sexual reproduction usually provides is missing. Note that genetic mutations here will only be expressed if they produce dominant alleles, assuming diploidy. Any recessive alleles would not constitute a reserve in that, in the absence of bisexual reproduction, there is no means of accessing them. (Of course, it is possible that both of a pair of genes might independently mutate into the same recessive allele in the same "family" line, but this seems improbable.) If the parthenogenetic individuals are haploid, all alleles will be expressed, but again, there can be no reserve of unexpressed alleles.

In parthenogenesis, every reproducing female produces a population of offspring that is independent, other than through ancestry, from similar "offspring populations" of other females. Rather than class each such population as a distinct species to be named (which would create a taxonomic nightmare from our finite perspective as humans), these populations are all lumped under one species name. They are recognized primarily by their morphological similarity, so are considered a "morphospecies."[33] Also, "paleospecies" that are morphologically distinct, extinct forms known only from fossils, which can hardly be tested for interbreeding ability, likewise fall in the category of "morphospecies."

Many sexually reproducing species, especially if they are very small[34] or in some other way have not gotten adequate attention from biologists to document their breeding patterns, are classed into species based on their morphology and hence, at least temporarily, are morphospecies in concept.

Sometimes two separately interbreeding, sympatric (occurring at the same geographic location), populations are nearly identical morphologically. Invariably, such populations have distinct ecological niches (according to the "competitive exclusion principle" explained in ecology texts) and at least certain DNA distinctions. When recognized, these groups are termed "sibling species," certainly fitting the biospecies concept. (The term "sibling" recognizes their comparatively recent common ancestry as evidenced by their lack of morphological divergence.)

Populations that have diverged, such as the satellite populations we discussed earlier, but retain interbreeding potential with each other at least somewhere in their geographic distributions, are usually considered races or "subspecies." The implication here is that these populations represent incipient, distinct biospecies, or at least have that potential. How they can achieve that status will be considered shortly. First, however, we must recognize one more concept of morphospecies.

With the advent of computers, a novel approach to taxonomy has occurred. *Cladistics*, as it is known, is based on sets of usually morphological traits of specimen samples. These work out on "cladograms" into final categories that are generally termed "phylogenetic species." Their relationship to ecological niches, as well as their degree of reproductive isolation, is generally omitted in their conceptualization. Hence, this, too, is a morphospecies concept.

This concludes our concept setting for the meaning of the term "species." Now let us consider the manner in which new biospecies arise. What is the process of speciation and how does it fit the biblical perspective?

Part III: Science Reveals an Allegorical Meaning of the Biblical Account of Speciation

Psalm 104:29-30, clearly states that God (as his immanent Spirit) replaces "species" with new ones and that this involves their ecology (104:27: "they all look to you to give them their food at the proper time." And 104:29: "when you hide your face . . ., when you take away their breath, they die . . .") We do not expect that the Psalmist nor his contemporaries understood the hidden, allegorical meaning of this statement. But the modern science of population biology clearly reveals it!

We earlier stated that, from a functional aspect, the most significant of our species concepts is the biospecies, the concept based on reproductive isolation. That is the concept we deal with here. (We note here that the fate of a completely parthenogenetic species is eventual extinction concomitant with a loss of its niche. Without the reshuffling of genes allowed by normal sexual reproduction, adequate adaptive changes are improbable.) We have seen how directional selection can shift the adaptive peaks of sexually reproducing populations. Documented examples of the effectiveness of this process are well-represented in breeding programs of domesticated organisms. There are highly divergent breeds of dogs, for instance. But they are still all "dogs" of the same biospecies, *Canis familiaris*. Yet coyotes and wolves, hardly more different in morphology than some breeds of dogs, are separate biospecies, *Canis latrans* and *Canis occidentalis* respectively (versus *C. lupus* of Europe and other wolf species). Obviously, the difference in these circumstances involves the degree of gene exchange ("flow"), which is dependent upon interbreeding. Even though the individuals making up two populations are similar, if they do not interbreed in nature, that is, they have a "reproductive barrier," they are separate biospecies and have the potential for independent divergence.

But just what constitutes a reproductive barrier? If dogs of extremely different breeds such as Great Danes and Pomeranians were the only survivors, the limits to their interbreeding ability is obvious. However, as long as there are existing breeds of intermediate size, there is at least the potential of natural gene flow, so that they are not considered separate biospecies. This is not to say that if they were only known as fossils, or were dealt with only in cladograms, that they would not be classed as separate paleospecies or phylogenetic species respectively. Such is the "confusion" inherent in our attempts to categorize nature from our finite human perspective!

But why don't wolves and coyotes interbreed? It is not likely due to their size difference. Taking a clue from the Psalmist, we should consider their ecologies. We see that wolves and coyotes have different niches, preying upon different categories of animals. Presumably, a hybrid between them would fit neither niche and therefore, be at a selective disadvantage. So what keeps wolves and coyotes from interbreeding in nature? Obviously, those who don't misbreed will leave more viable offspring.

Let us again turn to the Bible for an answer. In Genesis 6:2, we read "the sons of God saw that the daughters of men were beautiful, and they married any of them they chose." Genesis 6:4 calls the apparent hybrids, "Nephilim," and implies that they were characterized by what biologists would term "hybrid vigor."

We will return later, in a different context, to another aspect of the significance of this statement. But here note that interbreeding was dependent upon sexual attraction. Suppose that the sons of God had only found the daughters of men to be repulsively ugly. Obviously, they would not have been motivated to marry them and, therefore, no offspring would have been produced. We see here the importance of mate selection as a reproductive barrier. Presumably wolves find coyotes "ugly" and vice versa. Directional selection would create such a reproductive barrier when those individuals that find the opposing population's individuals sexually repulsive leave more offspring than those that do misbreed. Eventually, genes controlling sexual behavior, which promotes misbreeding, would be lost from the gene pool as the hybrids die off without leaving offspring. We biologists declare that, once this process is concluded, speciation has occurred.

We must recognize that there are other routes to the creation of reproductive barriers. For instance, at least as far as we know, plants are not able to appreciate each other's beauty or ugliness, but if diverging populations were, for instance, to flower at different seasons, or attract different pollinators, reproductive barriers would result.

We recognize here that there may still be varying degrees of genetic compatibility between the new biospecies. My colleagues and I have done research in this area that is soon to be published. But note here that it is the ecological niche compatibility of each population that is served by the reproductive barrier. The Psalmist was right even though the immediate meaning of his allegorical statement was all that he could understand!

We note also that, in addition to God's allele mutation "storm" (like his rainstorm) yielding the needed traits to actualize such speciation events, God also determines the "genetic needs" boundaries by his control of the parameters of ecological niches (Psalm 104:27-28).

And thus God said, "let the land produce living creatures according to their kinds" (Genesis 1:24). Note that every kind of creature is intended by God and declared to be good. And how does humankind fit into this context? First, let us again look to the Psalmist, here at Psalms 139:13-16. The implication here is that God is in control of the embryonic development of each of us as he was with the Psalmist personally ("I was made in the secret place"). Thus, indeed, we are each personally created by God. But we also see that the actualization of God's creative intentions involved his guidance of our long term ancestry and were even programmed and guided through cosmological evolution in preparation for our appearance ("I was woven together in the depths of the Earth"). Of course, there is an apparent meaning, if this is allegory, although the words used," depths of the Earth," seem strange for a mother's womb. However, these words nicely fit the hidden meaning here proposed, especially if the term "Earth" is taken to denote the cosmos as a whole.[35]

Part IV: Adam and Eve

This brings us to the question: who were the biblical Adam and Eve? How can the biblical account of them relate to the allegorical meanings revealed by modern science such as we have discussed above? One thing we may note is that the appearance (creation) of humans in the biblical account is preceded by the creation of all the other life forms. This fits the modern position given *Homo sapiens* in relation to our ancestral life forms.

Dora Jane Hamblin in her online posting, "Has the Garden of Eden been located at last?" (http://www.ldolphin.org/eden/), provides in-depth evidence of the historicity of the Genesis account. For instance, she cites the work of Dr. Juris Zarins of Southwest Missouri State University in Springfield, who holds that the Garden of Eden lies presently under the waters of the Persian Gulf. Quoting Hamblin: "he further believes that the story of Adam and Eve in—and especially out—of the Garden is a highly condensed and evocative account of perhaps the greatest revolution that ever shook mankind: the shift from hunting-gathering to agriculture." In his extensive investigation, Dr. Zarins ties the written biblical account to apparent word-of-mouth historical accounts passed down through the generations of the ancients. Thus, the biblical account relates to the stories, recorded elsewhere, from as far back as the prehistoric Ubaidians, or the "Sumerians who invented writing and the Assyrians who absorbed Sumer's writing as well as its legend of a luxuriantly lovely land, an Eden called Dilmun." Also involved were the Kashites in Mesopotamia, "contemporaries of the Israelites then forming the state of Israel."

Dora Hamblin suggests the significance of two dates. First, 30,000 BC, when modern humans replaced Neanderthals. The other, about 6,000 BC, when agriculturists (Ubaidians) displaced the more primitive hunter-gatherers. Her fascinating account not only documents these matters, but deals also with many more aspects including an account of Adam and Eve. For instance, her account of the Adam's rib origin of Eve is particularly informative. In our present context, we recognize the significance of these accounts in richly establishing the historic validity of Genesis.

But what allegorical significance does the Genesis account of Adam and Eve hold? We have seen profound hidden meaning, and thereby, evidence of inspiration from transcendent source, in the Genesis account of the physical and biological aspects of creation. What hidden meaning might there be in the biblical version of the creation of humans derived from this historical context? And wherein might such meaning evidence a transcendent source? And why are there two creation accounts in Genesis?

Let us make some assumptions and do some speculating to see if we can derive a sensible explanation and then consider the biblical evidence bearing on the matter. We will draw on several of our earlier insights.

Assuming that the ekpyrotic model of the universe is correct and that the brane opposing ours is real, could it be that the biblical Garden of Eden is actually in that other brane? If so, how would this perspective fit the Genesis account? First, we note that the biblical account leaves us with the assumption that the Garden of Eden still exists, at least there is no account or prediction of its demise. As we have seen earlier, apparently there were earthly sites known for their garden-like qualities. But it seems unlikely that they ever contained such plants as the "tree of the knowledge of good and evil" or the "tree of life," Certainly such plants would not fit any of the phyla of plants known even from fossils.

Interestingly, the biblical book of Revelation holds that at least the tree of life continues its existence in "paradise" (heaven). (See Rev. 2:7 and 22:2-14.) Should we assume that it has somehow been transplanted, or is it still where it has always been?

On the other hand, Genesis is quite clear on the matter that Adam and Eve were "transplanted" upon their disobedience to God's command. We note that the writer of Genesis would have had no concept of extra-spatial dimensions, so how might they be described in his account? Note that God planted the garden in "the east" (Genesis 2:8). Note that Adam and Eve were expelled, apparently to the east because it was here that "the" cherubim with a "flashing sword" was (is?) stationed (Genesis 3:29). (Might these "flashings" or light generations, denote passage through the fourth spatial dimension as we have earlier considered?) Further note that in Psalms 103:11-12, the distance between "heaven and Earth" is compared to the distance "between" the "east and west." Now if "heaven and Earth" are taken to allegorically refer to the two branes of the ekpyrotic model, and their separation distance (in the fourth dimension) is equivalent to the distance "between the east and the west," then it would seem reasonable to conclude that the ancient writer's use of the term "east" to denote displacement could be taken to allegorically refer to the fourth dimension separating the ekpyrotic branes. If these reasonings are correct, the implications are profound.

For instance, if the law of entropy is somehow different in the "heaven" brane, we see how it is that Adam and Eve, and perhaps the "animals" of paradise, would not die. Furthermore, we can relate what I call the "Jonah principle" here. That is, God deals with disobedience, which is a challenge to God's character, by banishment, Jonah to the "great fish" that the "Lord provided" in the ocean and Adam and Eve perhaps to an earthly mortal primate, likewise provided by God, within this entropic brane (Jonah 1:17).The struggle for survival characteristic of earthly ecosystems is well symbolized by the plight of toil to which Adam is condemned.

As with Jonah who, upon his repentance and subsequent obedience, was released from the fish in the ocean and returned to land, so, the Bible holds, God has provided a way to return the repentant offspring of Adam to paradise. Note, too, the symbolism of the events at Calvary during Christ's crucifixion. Only one of the two lawbreakers crucified with Christ repented and believed, in obedience to Christ. Jesus then told him "today you will be with me in paradise" (Luke 23:43). The Jonah principle: Disobedience results in banishment, repentance, and subsequent obedience results in restoration. However, all disobedience must be punished if God's perfect character is to be preserved. Hence, Christ's sacrifice which has provided the way for our (and Jonah's) reconciliation with God.

But why should transcendent God care about us in the plight earned us by disobedience? John 3:16 tells us that God loves us and promises eternal life to all who obey his command to believe in his son, Jesus Christ. We have seen that, according to the Bible and our earlier reasoning, God is infinite. This must mean in all dimensions, including those of time. God is thus "eternal" as infinitely complex, absolute somethingness which exists, according to our earlier reasoning, because it cannot not exist. In Titus 1:2, we read "hope of eternal life, which God, who does not lie, promised before the beginning of time . . ." In Psalms 139:16, we read "all the days ordained for me were written in your book before one of them came to be." Psalms 139:17 continues "how precious to me are your thoughts, O God!" The Bible holds that God is not only eternal but also immutable. For instance, such is the meaning of the statement made to Moses in Exodus 3:14 at the "burning bush" event. "I AM WHO I AM." Might it be, then, that each of us has had an eternal potential "in God's mind"?

What, then, is the biblical perspective of our place in existence? Suppose that we do have an eternal potential for existence in God's mind. In God's will, this is to involve sentience and free choice to honor God's perfect character by obedience to his commands. God first actualizes mankind's potential in Adam and Eve and places them in the context of the Garden of Eden, or Paradise, within the heaven brane. Eve and Adam challenge God's character by exercising their free will in disobedience. They are expelled from Eden to the "east," apparently meaning through the fourth spatial dimension into the earth brane. Here God has prepared recipients for their "sentience," within a context of earthly life forms. (Recall that God regards all the life forms he has created, including all the species of primates, as "good." See the Genesis, chapter one account.) God's preparation of these earthly recipients, we have reasoned, has involved modification of life-forms, eventually producing the recipient species, *Homo sapiens* (Psalms 139:15). The first creation account given in Genesis, chapter one, could be taken to denote this aspect. The second creation account, dealing more specifically with Adam and Eve, is given in chapter two and can be taken to denote the events occurring within the heaven brane.

But what was the means of the transfer of "sentience" to primitive *Homo sapiens*? Genesis, chapter six, relates the crossbreeding of "the sons of God" with the "daughters of men." Might this be the means whereby such "sentience transfer" occurred? If so, then Adam and Eve, in their paradise brane form, must have been physically compatible with our earthly form. Would this mean that there had been a parallel sequence of life-form modification in the paradise brane? And what of the problem of dimensionality as it relates to the atomic structure needed to allow passage through the fourth dimension? (See earlier discussion.) Rather, it seems more likely that God transfers his actualization of our existence potential, eternal in his "mind," directly to each of our physical bodies within this entropic brane as compensation to his primary plan which was disrupted by Adam and Eve's disobedience. This may well involve emergence as God directs our embryonic development (Psalms 139:13) and subsequent events in our lives.

The biblical account in Genesis suggests that this "transfer," indeed, began with one man and one woman, "Adam and Eve." Note, however, that this would not involve total genetic uniqueness, but rather the emergence of sentience. In their earthly bodies, they were able to reproduce, perpetuating sentience potential to their descendants. Perhaps the biblical "sons of God" refers to these descendants.[36] That there were "daughters of men" with whom they could intermarry suggests that there were, indeed, other *Homo sapiens* around.

It would thus seem that our individual eternal existence potentials must be actualized or God's will is compromised. The exercise of Adam and Eve's free will to disobey necessitates a compensatory means of fulfilment. Thus, we are related to Adam and Eve in identity of our roles in God's act of actualizing eternal potentials. However, we find ourselves in the context of an entropic brane and inheriting Adam and Eve's fallen nature and thus deserving of eternal banishment. If we ever are to fulfil God's plan of free choice obedience to his will, our debt of punishment inherited from Adam and Eve and evidenced in our fallen nature, must first be paid. God's incarnation in Jesus and his sacrifice for our sins, the outcome of our fallen nature, pays this debt. Reconciliation to God's plan is thus a gift which we must choose to receive. Upon doing so, our nature is changed. We are "saved" from the earned fate of the disobedient, the fate of eternal banishment. We now have a changed identity. We are responsive to God's immanent Spirit. It is now "natural" to us to enjoy obedience to God's will in our lives (e.g., see 1 John 5) This is quite distinct from the learning of proper social behavior, although there may be similarities in behavioral expression. In that our sentience yet resides in mortal bodies, there is often conflict in motivation. St. Paul describes such conflict in Romans 7:14 to 25. In this context, "good" actions are the outcome of a regenerated nature, a "spiritual rebirth." Other religions do not hold to this premise and must, then, be regarded as human attempts to reach God or at least achieve social conformity rather than the God's interaction with us at our level.

As with the repentant thief next to Christ on a cross, we, too, will be restored to paradise if we thus obey and believe in Christ and his provision for redemption from our inherited fallen condition. Such is the core of the Bible's message.

The significance of establishing credibility to the Bible is now evident. Is it, indeed, communication from transcendent sentience? Only if so, can its message be true.

Part V: Loose Ends

Let us now take an overview of "cosmic emergence" from the perspective we have suggested above and deal with any "loose ends."

Absolute somethingness exists because it cannot not exist. This includes infinite sentience, God, who likewise exists because he cannot not exist. Eternal, immutable God, whose infinite sentience includes, perhaps in addition to those of other "finite sentiences," the eternal potential for our existence as sentient beings with free will.

In preparation of a context for our existence, God creates the relative somethingness of the present cycle of the ekpyrotic model, beginning with the "flat splat" that results in the creation of the present conditions in both the heaven brane and our earth branc. God first actualizes mankind's potential in Adam and Eve in the context of the paradise of the Garden of Eden that we have considered yet existent within the heaven brane. Adam and Eve exercise their free will in disobedience to God, thus challenging God's character.

Meanwhile, in our earth brane, God creates, through his natural processes, conditions suitable for the emergence of life. There is presently a lack of conclusive scientific evidence as to just how the complex molecular organization of even simple living cells first took place. However, I, for one, take confidence in that, no matter how improbable an event may seem from our finite perspective, it is certainty if God intends it. Thus, without yet knowing the specifics of the origin of life event, I am inclined to agree with Charles Darwin's perspective on the matter. He states this view in his *The Origin of Species* as follows, "there is a grandeur in this view of life, with its several powers, having been originally breathed by the Creator into a few forms or into one . . ." Thus, we see that Darwin was, indeed, a creationist, at least in accounting for life's origin.

Even so, God's "breathing" life into an original "form" may well have been through natural process. For instance, God's control of the decoherence and collapse of the probability waves of the needed components at the right time and place and in the right arrangement would make the origin of life a certainty as God actualized his intentions.

Of course, Darwin's major contribution was in recognizing a natural mechanism for chronological modification (evolution) resulting in speciation. Unfortunately, he did not recognize the "rainstorm" principle of God's immanent involvement in natural process and God's infinite perspective. For instance, Darwin suggested that the occurrence of "wasteful" mutations, together with selective parameters that are "natural," at least left questionable the involvement of divine design. From the perspective here expounded, we see the conceptual fallacy of Darwin's position. Yet, he obviously was not an atheist and one wonders to what degree he would have agreed with present "Darwinists"!

We continue our overview. Life originates and new levels of complexity emerge through God's natural process. In time, primates, including *Homo sapiens*, emerge in God's process of actualizing his eternal intentions, we mortals being an alternate for the original Adam and Eve who failed God by their disobedience. God yet actualizes his eternal will through those of us who freely choose to yield our will to his. This we do by accepting the gift of reconciliation to God made available by incarnate God, Jesus Christ, in his provision for our redemption through his sacrifice of coming to our level on this tiny planet and yielding his "life" on the cross. Thus, we now are God's accurate "footprints," true reflections of his image.

The biblical account records Jesus's outcry from the cross: "my God, my God, why have you forsaken me?" (Matthew 27:46). Was this God revoking the incarnation, or somehow a disruption of the trinity? In some way, Christ vicariously takes the blame for the results of Adam and Eve's disobedience, the sins of humanity. Restitution to the pre-fall condition of obedient free will that is the actualization of God's eternal intention is offered as a gift to repentant believers. Those who are non-repentant retain their eternal nature but are separated from God's benevolent presence in "hell" along with other disobedient beings such as "fallen angels" or demons. (However, being omnipresent, God is, at least in some sense, present everywhere.)

The biblical account further records that Christ was resurrected. Earlier, we considered the significance of his ability, in his resurrected form, to appear in a closed room. The Bible records also that his grave clothes were left behind in the tomb in a manner again suggestive of extra-dimensionality. (Apparently, Christ's resurrected form required apparel of a different than earthly nature.) In John 20:17, Jesus tells Mary Magdalene of his imminent return, "to my Father and your Father, to my God and your God. "Jesus is now to return to God, the Father, who "forsook" him when he was on the cross. And God, his Father, is now also God, our Father, as obedient believers. And in our time, we, too, will go to him in paradise. If the contentions we hold to here are correct, this involves returning to the brane context of pre-disobedient Adam and Eve.

Part VI: So Why Does the Universe Exist?

Is, then, the emergence of *Homo sapiens* as recipient of Adam's disobedient sentience the purpose of our universe? We seem so isolated in its vastness. What about other sentience somewhere on other planets? What about God's will even for individual birds? Again, we must recognize the "rainstorm principle" even at this level. God has infinite perspective which would include unnumbered objectives even outside our "garden plot." Christ's declaration, recorded in John 14:6, states, "I am the way and the truth and the life. No one comes to the Father except through me." This certainly must involve Christ's divine nature, his "I am" nature. And if so, may it not be that he, in his divine aspect, is also the "way" for sentience beyond our planet? In an expanded context, perhaps Adam was not the only disobedient sentience. The Bible characterizes Satan and the demons as angels fallen by disobedience, but now without hope. Perhaps we, as Adam's race, are unique in

God's intentions. It seems, though, that we are hardly in a circumstance to make judgment on these matters. Let us rather take heart that God has provided a way, at least for us, as we are repentant and obedient in our sentience.

Validity of this Perspective?

Earlier in our considerations, we noted the inadequacy of finite human attempts to interact with an infinite, transcendent God. We asked the question, "Has the God revealed himself to us as finite, sentient beings on this tiny planet?" We suggested the relevance of order imposed on quantum potential disorder as constituting God's "footprint," that is, evidence of his existence. We now have examined the scriptures of one of the monotheistic religions for its validity, first in establishing its historicity, then in its use of allegory with meaning hidden to the respective authors, but now revealed, especially by our advances in science. We have considered first several examples from physical science, and then examples from biology, especially regarding the matter of God's role in nature as it applies to life-forms. Finally, we have considered the Christian perspective of the biblical message of the purpose of our species, *Homo sapiens*. It is evident that this latter concept is wholly dependent upon the biblical scriptures and its actuality upon their validity as revelation from the God. Perhaps the significance of our quest for valid unwitting allegory is now especially apparent.

In our next section, we shall consider more examples of allegory within a biological context. We will, at some length, look for evidences for chronological modification, primarily in Lepidoptera, my other group of specialization. But, let us first further delineate the context.

CHAPTER SIX:
The Biblical Chronological Modification Model Compared to the Instantaneously Complete Model of Creation

We already have considered the basics of our divinely guided, chronological modification or evolutionary model which holds that immanent God actualizes his transcendent intentions primarily through the ongoing exercise of his natural laws. The outcome implicit in this process is a biological dynamic concomitant with the environmental dynamic characteristic of our planet Earth. God, through his "natural" processes, continuously allows genetic change and eliminates genetic deterioration so that life-forms adapt well to their ecological circumstances. (Again see Psalm 104:27-30.) Thus, this model predicts change in life-forms which allows continuous fitness, even though environmental niches change, and also allows for the emergence of increasing complexity. One expects new organs, etc. to be derivatives of previous organs or structures. No completely novel, complex structures should be found independent of a history of chronological modification. However, the model would allow for exceptional events at the quantum level.

In contrast, the instantaneously complete creation models imply an original perfection of form resulting from God's creative action. These creation acts are usually held to have occurred during the Genesis "days" (taken literally), and with no allegorical meaning. This perspective is held especially by those promoting a young Earth creation model. A modification in which new, instantaneous, creation events are held to occur repeatedly at even long time-intervals is held especially by certain proponents of the old Earth creation model perspective. Because avoiding "straw man" models is risky, let us here let these groups speak for themselves as to the implications of these basic tenets. Thus, when investigating their claims, one should look for their position on the following matters:

First, the basis for the young Earth biblical perspective apparently is primarily the conclusion that the Genesis creation account is literal only and without allegoric meaning. Thus, God's actualization of his intentions are not by natural process. Much of their rhetoric involves negative criticism of the tenets of natural science (often a "straw man" version). Thus, their version of God's creative action is promoted by default rather than by positive argument. However, the implications of this position concede to the naturalism perspective at least two critical points: God does not act through natural law and supraquantum chance is real. Rather, it is held that God has acted creatively only at an initial stage at which time the created beings were perfect. Given these concessions, it would follow that change through time brings only deterioration such as in the fitness of species. For instance, the races of humankind would not be considered each best fit to its natural environment, but rather at varying degrees of deterioration from the original, perfect condition at the supposed creation act. Also, it follows that each of us is not an individual creation by God, but rather a degraded "edition" of the original created condition. Further, it seems difficult to reconcile this position with other Bible passages, for instance Psalm 139: 13, 139:15-16. Also, denying the tenets of the modern sciences of astronomy and cosmology eliminates acknowledgement of such perspectives as the big bang and ekpyrotic models. Evidence of biblical inspiration through allegory also is forfeited.

In that species are not related by descent in this model, it would seem that completely novel structures should occur regularly independent of a history of chronological modification. For instance, porcupine quills should be distinct structures unrelated to hairs, and without intergrades. (However, such intergrades do occur even on the same porcupine specimens. See figure 5.) Structures, such as angel wings with novel bone structure, perhaps even with feathers, would be found, for instance, in mammals. And why shouldn't bats' wings be provided with flight-efficient feathers such as found in birds?

Early versions of the young Earth, non-allegorical perspective held that the planet Earth is flat, not spherical. Current versions that deny modern cosmological models forfeit the allegorical meaning of the biblical statements which suggest that the earth brane, our universe, is three-dimensionally flat, a fact recently confirmed by astronomers.

Certain old Earth versions, which, even so, hold that biological kinds were created complete instantaneously, do accept astronomical and cosmological contribution, but, as do the young Earth versions, deny God's ongoing involvement in natural process, at least as it relates to biological evolution.

I am unaware of any of these perspectives that recognize a means whereby instantaneous creation may occur, other than by God-performed supernatural action (or from our perspective, by "magic"). A life form wasn't, then instantly, it was. However, if one accepts the current findings of the science of quantum mechanics, particularly Heisenberg's uncertainty principle, certain means of instantaneous creation may be possible. Because no new absolute somethingness can be added to infinite, absolute somethingness (otherwise, it would not have been infinite), we must start with something. The Bible suggests the "dust of the Earth." We now would say "matter."[37] The uncertainty principle holds that each unit of matter has a characteristic probability wave.[38] Until each probability wave decoheres and collapses into one of the possible realities, the character of the unit is not determined (Copenhagen interpretation). And apparently, the unit's probability wave nature is frequently regained and thereby new decoherence events can occur with regularity. We have earlier suggested that God, in his immanency, determines into which reality each probability wave will collapse.[39] (See also endnote 9.)

Suppose, then, that the required amount and kinds of matter, perhaps scattered, has simultaneous generation of probability waves of all of its units and that the probability wave of each unit then decoheres, each in the required position, into the composite reality needed to make up a living organism. Ordinarily, this would seem highly improbable. But would it not be at least possible, even within natural quantum law? We earlier postulated such an event for the origin of the first living cell, following Charles Darwin's suggestion of such an initial creation act of God. And perhaps it is the means whereby miraculous healings occur.

But where is the evidence that God has used this process in his general creative action? And, if he hasn't, at least it would seem more probable, according to God's natural law, that gene mutations needed for an increase in organizational level of a phylogenetic line of organisms would occur as part of God's ongoing creative action, keeping populations adapted to their dynamic ecological context. One should realize that God, being eternal, certainly does not have chronophobia! Why then, should there be any disadvantage to chronological modification as God's creative mode?

Studies in ecology evidence the actuality of the Psalmist's declaration of the interdependence of organisms with their environment (again, Psalm 104). Thus, creation must involve not only the animal or plant form, but also a suitable environment with which it interacts. And interaction requires time. Thus, acknowledging God's creation of dynamic ecosystems through a process of evolution is awe-inspiring. When understood and properly revered, indeed, it is an act of worship.

CHAPTER SEVEN:
Scientific Evidence Supporting the Chronological Modification Creation Model

The science of biological evolution holds the following basic tenets: Changes in populations of organisms occur in response to directional selection. This involves gene frequency shifts. In inter-breeding organisms, this is enhanced by the shuffling of the alleles already in the gene pool, such shuffling allowed by sexual reproduction. There is also the eventual addition of new alleles through mutation and, perhaps, even by viral transport from other species as apparently evidenced by recent genome research. Speciation occurs when interbreeding populations become reproductively isolated from each other. Directional selection may channel each such population into an unique ecological niche, and when reproductive barriers form, each population is allowed an independent course of further change. As independent lineages progress in changing through extended time periods, significant differences in their characteristics are realized. These patterns are represented by biologists as phylogenetic trees. A classification system delineating the degrees of divergence in supra species levels involves categories such as the genus, family, order, class, phylum, and king-dom, in order of increasing inclusiveness and degree of distinction.

However, as was pointed out earlier, the basic unit in this system is the biospecies. Whether or not interbreeding populations are reproductively isolated is testable, at least at a present time horizon and under suitable circumstances. Indirect evidence of such reproductive isolation may involve morphological distinction taken to be a consequence of the lack of sharing mutational alterations novel to the isolated status.

According to this model, in that speciation is a dynamic process, it is considered to involve sev-eral stages. At a given time horizon, it is to be expected that different lineages will be at different stages in this process. Thus, although it would be difficult for a scientist to follow the process to completion within a given lineage due to time constraints, finding examples of the several stages at a present time horizon, but within different lineages should be possible within the pertinent con-text. Figure 6 illustrates this situation.

One recognizes that these stages vary in length, stages one and five being much longer lasting. Thus, examples of these two stages should be, and are, much more numerous. I will cite a few ex-amples before concentrating on stages two and three.

Any species showing geographic variation of sufficient degree illustrates stage one. Our own species, *Homo sapiens*, illustrates such variation sufficiently to at least establish races, each adapted well to the environment in which it has developed. For instance, humans adapted to areas near the equator have denser skin pigmentation, giving better protection from the more intense so-lar radiation occurring there. Among insects, the principle is well illustrated by the Papilionid but-terfly, *Papilio indra*, found in western North America. Some recognized subspecies are illustrated and named in figure 7. A map of the geographic distribution of these subspecies is given in figure 8. In such situations, continued gene flow between the subspecies, as evidenced, for instance, by populations of intermediate characteristics, at least potentially retains the whole as one naturally interbreeding population which thus is considered a biospecies.

Stage five is well-illustrated by certain other species grouped with *P. indra* in the same genus. Apparently, these no longer interbreed in nature. Here we choose the following species as representatives, *P. eurymedon* and *P. rutulus*, illustrated in figure 9.

I have chosen the Papilionid species, *Parnassius smintheus*, to illustrate stage two. *P. smintheus sternitzkyi* is presently geographically isolated in the Siskiyou Mountains of southwest Oregon and into adjacent California, while *P. smintheus olympianna* is isolated in the Olympic Mountains in Washington. These populations have apparently been isolated from each other, as well as from the Cascade Mountains and eastward populations of the species at least since the retreat of the glaciers of the last ice age, the Wisconsin, some 18,000 to 10,000 years ago. Of course, some degree of divergence may have occurred prior to this isolation. All of the populations use plants in the genus *Sedum* (Crassulaceae) as larval hosts. At least *sternitzkyi* has occupied a unique ecological niche in that it uses *Sedum oregonense* exclusively as its larval host. *P. s. olympianna* uses *S. divergens* (confirmed by my personal observation during June of 2009). Other subspecies use yet other Sedum species.

Adaptation to a specific plant host may well involve chemical factors if they are unique to the given host plant. Also, behavioral adjustments allowing proper timing to coincide with the availability of new leaf growth, etc. and perhaps even resting site location selection for time spent between larval feedings may be involved in such adaptation to the extent that these factors are genetically controlled. Figure 10 illustrates the pertinent subspecies and shows the only other species of this genus found in the area, *Parnassius clodius*, for comparison. The larvae of this species feed on wild bleeding heart, *Dicentra formosa*, and other species in the family Fumariaceae.

Apparently, no laboratory crossbreeding experiments have yet been done with the various subspecies of *P. smintheus*. The results, especially of an attempted cross of *P. s. sternitzkyi* with any of the others would be of interest in our present context. Even so, what matters most in our context of the process of speciation is what would happen if these isolated populations were to re-establish contact. For instance, this may be allowed in a future ice age in which the pertinent habitat would likely retreat to lower elevations and hence have the potential of extending its geographic range. Would the respective populations develop reproductive barriers allowing their distinctions to persist, resulting in independent evolutionary futures as then distinct species? This would constitute stage three in our model.

Examples of Stage Three: The Point of Speciation

But what presently constitutes a barrier to mountain populations is rather, for lowland populations, a corridor for expansion. Such populations, isolated during the last glacial period into refugia, are now with extended range so that they may make contact, giving opportunity to interbreed. Here we could witness stage three of our model.

Among butterfly species fitting this stage is the swallowtail, *Papilio machaon* (Papilionidae) illustrated in figure 11. This species, actually a species complex, has been studied extensively by several biologists. Among them are C. A. Clarke and P. M. Sheppard who, on page 199 of vol. IX of *Evolution* (June, 1955), state the following: "It is clear that the machaon group provides some of the most suitable material ever investigated in animals for studying the process of speciation in detail, taking into account genetic, ecological, and behavior differences as well as time." In this paper, "A Preliminary Report on the Genetics of the Machaon Group of Swallowtail Butterflies," Clarke and Sheppard compare the genetics of four species in this complex: *Papilio machaon, P. zelicaon, P. polyxenes*, and *P. brevicauda*.

More recently, other authors, such as Felix Sperling and Wayne Wehling, have studied the western North American representatives in considerable depth, especially in regard to their population ecology and geographic variation. Later, we will examine their results as they are pertinent to our considerations. However, let us first better set the context.

Perhaps it should not be surprising that the "point of speciation" (stage three in our model), is complicated. We have suggested a time line in the process. However, it must be recognized that there is a spatial component at any given time horizon. Hence, in a sense, every spatial position in an extended population's distribution has its own time line.[40] This both complicates our considerations but also provides opportunity for detailed comparison of the stages represented by the spatially distributed time lines, all done at our present time horizon. In our considerations of the example provided by this well-studied complex, we will attempt to establish principles of general pertinence.

There are several "units" within the machaon complex spread over both Eurasia, North America, and one extension into South America (*P. polyxenes americus*). Because our purpose here is to establish principles, we needn't explore all the details of distribution and divergence within the complex, which would take at least one book in itself. Rather, we will concentrate on one area of the complex, those forms found in mostly western North America. This area is chosen because I have the most personal experience with these populations. But we do need some consideration of their ancestral forms. Here it is convenient that, among insects at least, ancestral species often survive to be contemporary with their daughter species which, as a rule, occupy different ecological niches.

Because more "primitive" species of swallowtail butterflies, such as *Papilio alexanor* (see fig. 12), are found in Eurasia, it is concluded that the Eurasian *P. machaon* is the stem form of the machaon complex. Here the degree of primitiveness is judged on the basis of generalization of traits. For instance, *P. alexanor* has a wing pattern as an adult similar to that of the tiger swallowtails, e.g. *P. rutulus* (figs. 9,12) and its related species, but has larval characteristics similar to the machaon group (see fig. 13) and uses Apiaceae (carrot family) species as hosts, as do most machaon populations. It apparently is found characteristically in alpine meadows within its range.[41]

Thus, early *P. alexanor* is considered to have given rise to the Eurasian machaon component, which in turn has given and continues to give rise to the other forms in the machaon complex. Felix Sperling and Paul Feeny have shown that the present population of European machaon and the North American *P. bairdii* are genetically close (0.2% divergence in mitochondrial DNA) and on this basis consider them conspecific.[42] This is based primarily on their genetic affinity, but overlooks the fact that they apparently do not presently interbreed. *P. bairdii* (including *P. b. oregonius*) had apparently been isolated to the south from the northern machaon subspecies *P. m. aliaska* and *P. m. hudsonianus* (fig. 11) during the last glacial advance. However, recently during the hypsithermal period with the retreat of the glacial barriers, *P. bairdii* had extended its range (as the subspecies *oregonius*), so that it interfaced with *P. machaon aliaska* in the Peace River Valley of British Columbia and Alberta, Canada. At present, populations of the subspecies *P. b. oregonius* are disjunct to the south from *aliaska*. However, the population known as *pikei* (fig. 11d), which is of questionable origin as will be discussed later, is presently at least parapatric to *aliaska* populations in the Peace River drainage. And there is little evidence that these populations presently interbreed in nature as indicated below.

We should note food-plant relationships in our considerations. In nature *Artemisia dracunculus* (Compositae) is apparently the exclusive host-plant of all of the subspecies of *P. bairdii*. However, according to Guppy and Shepard in *Butterflies of British Columbia*, [(Vancouver, Toronto: UBC Press, 2001), 130], in laboratory trials, *pikei* does better feeding on *Heraculum maximum* (Apiaceae), but will also feed and mature on *Artemisia dracunculus*. Also, I have found that larvae of any of the subspecies of *bairdii* will accept the introduced *Foeniculum vulgare* (fennel, family Apiaceae) as a laboratory food-plant, although females only rarely can be induced to use it for oviposition. As stated, in nature, *A. dracunculus* is the exclusive host plant of all of the populations of the *P. bairdii* subspecies to the south. The significance of this will be discussed later in the context of the niche distinctions of sympatric populations of the macaon complex species, *P. bairdii*, *P. zelicaon*, and *P. indra*.

In British Columbia, *P. m. aliaska* uses *Artemisia norvegica* (mountain sagewort) as a host (Guppy and Shepard, ibid., 131) as well as *A. arctica* and occasional use of Apiaceae such as *Conioselinum dawsoni*.[42] This is in contrast to machaon's exclusive use of Apiaceae hosts in Europe. (In parts of Asia, machaon populations apparently may use either Apiaceae or various Composite family, *Artemisia* species as hosts.)

We should note here that *Papilio bairdii dodi* is a subspecies found mostly in southern Alberta where it apparently is allopatric to *P. machaon aliaska* but apparently is at least parapatric if not sympatric to populations of *P. m. hudsonianus* with which it apparently does not hybridize. (See the distribution maps in Guppy and Shepard, ibid., 129-131.)

We now come to the focal point of stage three in our speciation consideration. Quoting from Guppy and Shepard, on page 131: "the *P. machaon* subspecies rarely hybridize with the *P. bairdii* subspecies *dodi* and *pikei*, even when they are sympatric or near sympatric, indicating reproductive isolation." They speculate that the Peace River lowlands were colonized by *P. bairdii oregonius* during the warm hypsithermal period just after the last ice age and that this species may have had limited hybridization with *aliaska* at that time, resulting in enough gene exchange to delineate *pikei* as a distinct subspecies of *bairdii*. Quoting again from the same page, Guppy and Shepard state, "the genotype of *pikei* is now stable, lacking hybridization with nearby *aliaska* populations, which indicates reproductive isolation and hence speciation." And this is in spite of the close identity of their genotypes as indicated above (machaon vs. bairdii). It thus appears that stage three of our model is presently occurring here, having begun at the geologically recent hypsithermal period and is now coming to completion at our time horizon.

But this is not the end of the matter. It turns out that the related, but more genetically distant species (1.25% mitochondral DNA), *P. zelicaon* in fact does hybridize with *P. machaon aliaska*.[43] The machaon subspecies "*hudsonianus* and possibly *aliaska* freely hybridize with *P. zelicaon* . . . " (Ibid., 131). One area where this occurs is along the eastern front range of the Rocky Mountains in Alberta, Canada. Another is at Pink Mountain, British Columbia. Figure 11 illustrates specimens from these areas that show both *aliaska* and *zelicaon* traits. Note especially the condition of the "eye spot" and its black border on the hind wings. In *P. zelicaon* a well-formed, central dot occurs, apparently adaptively derived from the original black border as found in Eurasian *P. machaon* and persisting in most *P. bairdii* populations in North America. (The distinct black "pupil" of the zelicaon eyespot apparently serves more effectively as a decoy to predator attack.) Note that the illustrated specimens from Nordegg, Alberta (Hwy. 11, west of Rocky Mt. House) and from Pink Mountain have an intermediate condition of the black eyespot. And compare these wild specimens with the laboratory-bred hybrid specimen resulting from the forced mating of a male *P.m. aliaska* from Pink Mountain, B.C., Canada with a female *P. zelicaon* from Oregon (all represented in fig. 11). Stock of the natural hybrid population from Nordegg was taken at 5000' elevation where its larvae feed on a species of *Heraculum* (Apiaceae). Thus, we see, its ecological niche is close to that of *P. machaon*, high elevation on an Apiaceae host rather than low elevation on the Composite, *Artemisia dracunculus*, that characterizes the niche of *P. bairdii dodi*. (See the specimens from Cut Bank, Montana, fig. 14.) The hybrid's niche is also close to that characterizing *P. zelicaon*, which, however, often is also found at low elevations.

Thus, it appears that there is here a paradox. The more genetically similar *P. machaon* and *P. bairdii* qualify, by their reproductive isolation, as distinct biospecies whereas the more genetically distant *P. machaon* and *P. zelicaon*, because of their natural hybridization, qualify as the same species by the biospecies concept. However, it must be recognized that some sympatric populations of *P. zelicaon* and *P. machaon* do not hybridize. In this situation, they qualify functionally as separate species, although gene flow between them may still occur via the existing hybrid populations. And, we wonder, will the result eventually be reproductive isolation, or will they continue to hybridize indefinitely?

What can we learn about speciation from this situation? First, we observe the confusion in application of species vs. subspecies taxonomic names. Our conceptualization here depends upon our perspective of the dynamic situation. If we look at it from where the entities are coming, we are likely to lump the populations as conspecific. If we think of it as where they are or are about to be, we would likely taxonomically characterize them as distinct species. Various authors have taken one or the other approach.

Next, we may investigate the possible causes of what appears to be a discrepancy. To do this, let us further clarify the roles of these two major factors characterizing the inter-relationship of populations: ecological niche distinction and the degree of reproductive isolation.

The Roles of Reproductive Isolation and Niche Distinction

Perhaps the significance of reproductive isolation is already apparent. If gene exchange between sexually breeding populations has ceased, each has become independent of the other in future genetic alteration. Time and the degree of directional selection are both factors in the resulting degree of divergence. It would seem that eventually, descendent populations may become distinct at ever higher taxonomic categories: genus, family, order, etc. However, distinction at, for instance, the phylum level involves aspects of the basic body pattern which are not likely to be altered in descendent lineages. See Appendix I for further details.

Again, we must recognize that genetic change is driven by selection for niche adaptation (see chapter five, part I: Population Dynamics). Thus, the degree of niche distinction for each population is of primary consequence in determining the degree of genetic divergence. And if a well-adapted population's niche remains the same through time, it is expected, due to stabilizing selection, that it will, indeed, not diverge genetically, except perhaps in its gene pool of recessive, seldom-expressed, alleles. That is, alterations due to expressed genetic mutations that decrease the fitness of the population to its stable niche will be culled from the gene pool.

Suppose, though, that two populations become genetically isolated but remain in the same niche. Here the competitive exclusion principle holds that, if they are sympatric, one will become extinct.[44] In contrast, if the two genetically isolated populations adapt to separate niches, they may co-exist without competition. Further, in another aspect of this context, differences in directional selection due to niche modification in different geographic areas may result in genetic divergence at the subspecies level. (There may also be examples of subspeciation where populations at the same geographic location are sexually active at different seasons, at least in some invertebrates and perhaps plants.) Herein, then, is the significance of niche adaptation to the speciation process.

Niche Specialization in the *Papilio Machaon* Complex

Let us now consider the role of niche specialization in our example of speciation in the *Papilio machaon* complex. In doing so, we need to expand our context somewhat, considering some background factors that have produced the present situation.

Apparently, *Papilio machaon* itself represents the basal ancestor of the complex in Eurasia and *P. zelicaon* the basal ancestor form in the Americas. During periods of mutual geographic isolation, these two mega-populations have accumulated independent genetic changes which have involved niche shifting although they are both rather "niche generalists." The latter is illustrated by the observation that *P. machaon* utilizes host plants primarily in the Apiaceae (carrot family), but some places also uses certain Compositae (e.g., sages), and even Rutaceae (citrus). It is recognized that "many of the swallowtail food-plants are linked with one another by their common content of various classes of secondary (chemical) compounds." [See Paul Feeney,(Scriber, Tsubaki and Lederhouse editors) *Swallowtail Butterflies*, Scientific Publishers, Gainsville 1995):10] Even so, the utilization of each separate species of host would entail some degree of niche distinction.

In North America, *P. zelicaon* is extensively "polyphagous," utilizing numerous species of Apiaceae and recently colonizing citrus in much of California. But, notably, *zelicaon* proper apparently nowhere uses composite species as hosts nor will lab cultures accept it, at least in my experience. However, each of the hosts it does use may require characteristic adaptations such as chemical tolerance, ovipostition behavior, and flight period adjustments. Figure 15 documents the great variation in zelicaon's flight period concomitant with its use of different hosts in different geographic locations. Sperling and Feeny discuss the significance of the plant chemicals, furanocoumarins, both linear and angular, in relation to their effect on the suitability of the plants bearing them as hosts (Ibid., 300).

So What of the "Point" of Speciation?

What makes our example of the speciation process pertinent, especially to stage three in our proposed scheme, is that the effectiveness of the process, which may take considerable spans of time, is being naturally "tested" during our time horizon. Stebbins in his book, *Processes of Organic Evolution*, [(Englewood Cliffs: Prentice Hall, 1966), 95], calls this the "test of sympatry . . ." To pass this test, related populations must "retain their identity even though they live close enough together so that cross fertilization between them is possible . . ." If they pass this test, they are to be regarded as distinct biospecies.

We have seen that this natural test is being enacted in the Peace River area of western Canada where *P. machaon aliaska* and *pikei* are presently distinct with no hybridization and in southern Alberta where *P. m. hudsonianus* and *P. bairdii dodi* remain distinct with no hybridization.

The test also is in process along the east slope of the Rocky Mountains in Canada where *P. machaon hudsonians* and *P. zelicaon* do hybridize to varying degree and thus lose their genetic distinctions, at least in the hybrid zones. Questions come up here regarding the eventual outcome of this hybridization. Will it eventually yield to a reproductive barrier to the presently interacting populations? Might it eventually result in genetic homogeneity in expanded portions of the ranges of the populations? Or perhaps, even a new ice age will eventually again isolate the populations geographically.

But suppose, as in the case of zelicaon's interaction with machaon, that in some places the test is "passed," but in others it is not. The implication, then, is that the test is still in process. Our "stage three" is being illustrated.

There is another complicating factor which we should consider, especially affecting areas of interface (parapatry) of two interacting populations. For instance, take our example of *P. machaon aliaska* being well-adapted to the environment of higher elevations at northern latitudes and hybridizing with *P. zelicaon,* which is adapted to varying elevations at more southerly latitudes. As long as there is gene flow into the hybrid zone from both parent populations, one would expect any genetic changes affecting mating behavior to be swamped out by the genetic influx. Meanwhile, there apparently is gene exchange across the hybrid zone into the respective gene pools. But in this case, it obviously is not sufficient to eliminate the parent population distinctions which may well be maintained by stabilizing selection. Indeed, this hybrid zone has apparently persisted since the hypsithermal (altithermal) following the retreat of the continental glaciers which allowed interaction of the respective populations as they expanded their ranges. Perhaps biologists in future generations will learn the eventual outcome of this process. Now let us consider another possible speciation process, that resulting from a "founding propagule." Such a speciation process, indeed, may have been involved in our machaon complex examples.

Suppose that one or a very few individuals colonize a new area. For instance, consider the situation at the eastern end of the Bering land bridge on the American continent at the time that Asian machaon individuals entered and were to become bairdii. The earlier machaon forms that had already established, were apparently only able to use Apiaceaes as hosts, not having yet adapted to the use of composite host plants. The new immigrant machaon, now niche-changed to also use composites, encounter an area apparently devoid of Apiaceae and zelicaon. But there is an abundance of *Artemisia* growing in what were apparently dry flatlands. Now, if such a population were very small, either due to a small number of founding individuals or to a subsequent "bottleneck" in size, there would be much inbreeding. Suppose that the genetic trait for host selection for oviposition had two alleles. And suppose that females, heterozygous for these alleles, would oviposit on both Apiaceae and composite hosts, but individuals homozygous for Apiaceae use (AA) would not oviposit on composites, and those homozygous for composite use (CC) would only use these as hosts. Now if only composites were available, obviously AA individuals would not leave offspring and CC individuals would do well. AC heterozygotes would leave offspring, but one-fourth of them would be AA and not reproduce. It likely would not be long, especially in a small population, before the allele, A, would be eliminated, that is the allele, C, would become fixed. Admittedly, this is a simplified version in that there are likely several genes that effect oviposition behavior. But it illustrates a principle that might apply to all of the genes involved.

Now suppose that mate selection is determined by a gene, M, with a recessive allele, m, which when homozygous (mm), will only allow mating with other mm individuals. In a large population this likely would be a rare event and individuals with this genotype would be at a great disadvantage in finding a mate. However, in a small population where inbreeding is prevalent, so that brother-sister matings of Mm individuals would be common, the mm genotype would likely be frequent enough so that mm times mm matings would be common. As the allele, M, becomes rare, perhaps affected by genetic drift, MM and Mm individuals would be at a disadvantage in finding mates and the allele, m, might well become fixed. And this process likely would occur with many other genes for various traits as well.

Such a small founder population being in an unoccupied niche should likely soon expand its range. In the case of the population that eventually is to become bairdii, this expansion would eventually involve encounter with Apiaceae feeding zelicaon forms. Presumably, they would still have M fixed as their mating behavior allele. (Recall that they supposedly originated from Asian machaon prior to the mutational origin of the m allele.)

Such an eventual encounter would find zelicaon and the new bairdii reproductively isolated and in distinct ecological niches, zelicaon on Apiaceaes and bairdii on the Composite, *Artemisia sp.* And, indeed, this is presently the case. Furthermore, this scenario predicts the empirically validated situation with the more recently established machaon forms, *hudsonianus* and later *aliaska*. Zelicaon mates with aliaska, both still having the M mating allele. Bairdii does not mate with aliaska or hudsonianus, presumably having the m allele fixed. Undoubtedly, the actual situation is more complex than that here postulated. But perhaps this account will sufficiently illustrate the principles involved. As a hypothesis, it does predict certain outcomes that could be tested experimentally such as the involvement of an M allele.

Another pertinent situation is what is happening with recently introduced "invasive" species often considered pests. Daniel Simberloff of the University of Tennessee has a posting, which can be accessed on the Internet under "Google," titled, "The Role of Propagule Pressure in Biological Invasions." Here he states, "A continuing rain of propagules, particularly from a variety of sources, may erase or vitiate the expected genetic bottleneck for invasions initiated by a few individuals . . ., thereby enhancing likelihood of survival. For a few species, recent molecular evidence suggests ongoing propagule pressure aids an invasion to spread by introducing genetic variation adaptive for new areas and habitats. This phenomenon may also explain some time lags between establishment of a non-native species and its spread to become an invasive pest." This scenario is, of course, different in some respects than our example. However, it is current in its occurrence. Do any of these invasions constitute speciation events? Apparently, this is yet to be determined by researchers.

Let's consider one last observation on the matter. May it be that the population known as *pikei* is the result of a founder propagule? Has it developed reproductive isolation? Is it in a sufficiently distinctive niche to survive future competition as it may expand to encounter nearby machaon complex populations? Might it be a new species in the making? Again, future research will be needed to determine the outcome.

We have now completed our considerations on principles of the process of speciation.

CHAPTER EIGHT:
Niche Distinctions of the Machaon
Swallowtails of the Pacific Northwest

Psalm 104: 27-28: "These all look to you to give them their food at the proper time. When you give it to them, they gather it up; when you open your hand, they are satisfied with good things." (NIV, Zondervan)[45]

In this chapter, we focus on the dependence of creatures upon the resources in their environment, both food and other "good things" needed for their survival and adaptation. We recognize that it is "good" that a population's size does not exceed the resources that support it. This process, indeed, may be complex as we shall see.

Papilio zelicaon is the only one of three species in the machaon complex that occur in the Pacific Northwest to be found west of the Cascade Mountains in Washington and northern Oregon. Here it is widespread from the coast to the west and into the Cascades to the east including at higher elevations. Its larvae feed exclusively on various Apiaceae (carrot family) including introduced species such as fennel (wild anise), *Foeniculum vulgare*.

East of the Cascades, *P. zelicaon* is the most broadly distributed species of this complex. But in certain habitats, two other species are found as well: *P. bairdii oregonius* and *P. indra indra* (see figs. 14b, 7a). In our context, let us concentrate on an example of the latter area and see how it is that these related species coexist.

Papilio b. oregonius, the Oregon Swallowtail (Oregon's state insect), in nature oviposits exclusively on the Composite, *Artemisia dracunculus* (green sage), which grows profusely along the Columbia River and its tributaries east of the Cascades. Let us here concentrate our considerations on the Columbia River riparian zone as found east of The Dalles, and especially, on the Washington side of the river, in the vicinity of the north end of the Maryhill-Biggs Bridge (Highway 97) and below the Maryhill museum. Here this species has apparently coexisted with the zelicaon and indra swallowtails since before the last ice age. Here desert-like, rocky habitat extends well into the riparian zone so that it is common for its characteristic Apiaceae, *Lomatium grayi*, to grow in close proximity to *Artemisia dracunculus*. And both the zelicaon and indra swallowtails use this Apiaceae as a host plant for their larvae.

In that the Oregon swallowtail uses green sage exclusively and neither of the other two species do at all, certainly competition for its larval host plant resource is excluded. However, the adults may use the same flowers as a nectar source and all the life history stages are subject to apparently the same predators and parasitoids.[46] (Remember that a species' niche involves not only taking from, but also giving to, its ecosystem.)

As illustrated in figures 13c and 16, the larvae of *Papilio zelicaon* and *P. bairdii oregonius* are very similar in coloration and pattern. Females of both species oviposit on sun-exposed sprigs of their respective hosts. As evident in figure 17, the early instars (stages) of the larvae are black with a prominent dorsal white patch. Thus, their resemblance to a product of negative interest to potentially predaceous avian insectivores is obvious: bird droppings. Birds, along with reptiles, do not waste water by urinating. Rather, their nitrogenous wastes are further processed to form the white, solid, uric acid, rather than the water-soluble urea of urine producers. This uric acid is added to their fecal wastes, forming the characteristic white patch which these young machaon complex larvae mimic.

In the apparent absence of large birds in their micro-habitat, the larvae would lose their mimicry advantage as they grow to the larger size of especially the last instar. Now the larvae take on a pattern well-established in the machaon complex, even found, as previously noted, in the primitive *P. alexenor* (fig. 13b). The larvae are often well-exposed as they bask in sunlight. Even though their coloration is somewhat cryptic, still they are usually easy to spot, even by this human "predator" with eyesight much less acute then that of birds. However, as with other species of swallowtail butterflies, these larvae have an eversible, fleshy epidermal gland known as an *osmeterium* that is located mid-dorsally just behind the head. When disturbed, this forked structure is protruded toward the point of attack. It is bright orange in color and emits a strong, distinctive odor, both features apparently serving as warning signals to predators that have previously experienced its protective quality, the odor being effective even for those that are nocturnal. One can empirically determine this quality by touching one's tongue to a protruded osmeterium. Be prepared for a rather startling, sharp, and repulsive taste. A researcher, Keiichi Honda, has determined that there are at least five chemical compounds secreted by the osmeterium of the last larval instar of *P. machaon* (and other Papilio species). What is more, there is apparently a shift in secreted chemicals of this gland concomitant with the larval molt and color change as the larvae enter this last stage, the fifth instar.[47] (I have not yet tasted of the osmeterial secretions of instars before the fifth. It may be that they are more adapted to repulsing invertebrate predators that would target smaller prey.)

If, indeed, the osmeterial defense is effective and requires educating local predators, birds and perhaps lizards, then its use by the larvae of both the Oregon and Zelicaon swallowtails would, in fact, be "cooperative" rather than competitive to the extent that local predators are educated by both species to the butterflies' mutual benefit.

But where does the indra swallowtail fit in? We have noted that this species uses at least some of the same host plants as does *P. zelicaon*.[48]

The Ecological Role of *Papilio Indra*

In most of its geographic range where it coexists with *P. zelicaon*, *P. indra* oviposits on the lower leaves of the host plants it shares with zelicaon. Upon hatching, the early instar indra larvae typically perch on the underside of their leaf. Interestingly, the larvae in these indra populations lack the white, "uric acid," mimicking spots so that they are well-hidden in the dark shadows. Obviously, in this situation, they lack the warming advantage of sun exposure. However, apparently to compensate, females tend to oviposit on host plants that are growing by rocks sufficiently large so as to act as heat sinks. Also their activity is somewhat later in the season when the weather is warmer. But the larvae must complete their feeding and wander off to pupate before the late season sun dries up their host.

The larger, last instar larvae of indra are, with marking pattern and coloration, distinct from that of zelicaon and bairdii oregonius (fig. 13d). These last instar larvae hide in the shadow patterns at the base of their host plants, or nearby, and may even perch under loose rocks.

Because of its interesting geographic variations and difficulty of acquisition, this species is popular with some collectors. Once, I was in the field with a friend who had a special interest in the indra swallowtail. He was particularly adept at finding the larvae in the wild. This day, I discovered his secret. At one point he was around a corner from me and out of view. As I walked along the road, I heard a volley of loud shouting coming from my friend's direction. Upon hurriedly rounding the bend in the road, I found him hollering at a pile of rocks! Had he "jumped the rail" mentally, I wondered?

Yes, indeed, indra larvae do have a functional osmeterium. My friend, whom I perhaps wisely will not name, had discovered that even hidden larvae will evert these glands in response to sufficiently loud noise. Thus, after shouting at the rocks where they had refuge, he could then follow the scent they emitted and locate them. Although the color of indra's gland is less striking, it does share the characteristic odor with zelicaon and oregonius.

But then, one may ask, why hide if you are distasteful to predators? I am not aware of a clear answer. Perhaps their distastefulness is not sufficient to deter a very hungry, persistent predator so that being hidden gives added survival advantage. Even so, such niche behavior certainly is in contrast to that of *P. zelicaon*.

So then, we see that each of these three machaon complex species have distinct ecological niches and thus can coexist. On occasion, I have found larvae of all three species within a few feet of each other, but each restricted to its characteristic niche. Although oregonius larvae will feed on at least certain Apiaceae plants in the laboratory, in the field I have never found any that have left the Compositae green sage upon which their mother had deposited them as ova.

Of course, all of these three swallowtails give to the ecosystem beyond being food for predators. I have documented a complex of parasitoides in this study area. A species of "fairyflies" (Mymaridae: Hymenoptera) infests eggs of the swallowtails here. At about .65 mm in length, these multicelled, feather-winged wasps are smaller than some single-celled Protozoa. Numbers of them may mature within a single host egg which itself is only about 2 mm in diameter and nearly spherical. This fairyfly species is, as of this writing, apparently undescribed and thus without a scientific species name. The adult fairyflies eclose from the host egg by mid-spring. One wonders what these tiny insects do to survive until the following spring when their swallowtail hosts are again available. (There apparently are fairyflies of many species found around the Earth. All those known are parasitoids of insect eggs including even aquatic orders such as dragonflies: Odonata.)

At the Maryhill site, it appears that the major parasitoid of the machaon swallowtail larvae is another wasp, but in the family, Ichneumonidae. These attack their host larvae in the early stages, usually killing them by the third instar. At this site, both indra and zelicaon are single-brooded, their host plants drying up too soon for any later broods. However, in that the host green sage of *oregonius* lives up to its name and remains "green" all summer, this swallowtail has subsequent broods even into September.

In my studies at the Maryhill site, I kept records of the frequency of parasitization of *P. indra* and of *P. zelicaon* for comparison. Averaging the results for samples from over three years (1966, 1968, and 1970) gives these results. Of 164 indra larvae, 23% were parasitized. Of 31 zelicaon larvae, 77% were parasitized. Obviously the sun-basking positioning of the zelicaon larvae has its disadvantage as well as sun exposure advantage. (I only sampled zelicaon on the host it shares with indra: *Lomatium grayi*. In that it uses other hosts as well, the comparative local population is larger than my data show. Indra doesn't use these other hosts, likely because of their more upright growth pattern with a lack of lower, shaded leaves.)[49]

But the niche complexity is even deeper. The ichneuman wasp parasitoid here has a parasitoid species of its own, also a wasp in the family Ichneumonidae. Females of this hyperparasitoid must first locate a machaon complex caterpillar that already contains a primary parasitoid larva. Then she must insert her ovipositor in such a manner as to deposit an egg inside the caterpillar, and in association with the location of the primary larva, probably either inside it or attached to it. I know of no studies that show how it is that the hyperparasitoid can detect which caterpillars harbor a primary parasitoid larva, nor how they locate the larva inside the caterpillar's body.

At another site not far to the north, I recorded the following data: In 1971, of 17 zelicaon larvae sampled, 47% had primary parasitoid wasps. Of these, 12% had hyperparasitoids. Of 52 indra larvae sampled, 12% had primary parasitoids and of these, 16% were hyperparasitized.

And there is yet another parasitoid that attacks the later instars of machaon caterpillars. In this case, it is a fly in the family Tachinidae. At the Maryhill site, this fly is uncommon. However, I have found it in other populations of machaon complex larvae, such as to the north of Wenatchee, Washington, where it is likely the major parasitoid rather than the ichneuman wasp as found at Maryhill. I am not aware of hyperparasitoids attacking this fly species.

I have not studied which bird or lizard predators attack machaon complex larvae or adults at the Maryhill site. And there are likely vespid wasps, and even spiders, and lady beetles (Coccinellidae) that prey on them as well.

That their osmeterium may be an effective deterrent for at least the predaceous wasps is evidenced by an observation I made on a wasp attack of a larva of another species of swallowtail, *Papilio rutulus*. The account I recorded of that incident follows:

"At 1:15 p.m., standard time on September 7, 1995, I witnessed a most interesting incident. I was watching a late fourth instar *Papilio* (*Pterourus*) *rutulus* larva repairing its silken hammock on a quaking aspen leaf in my yard at Monmouth, Oregon. To do this, it had turned from its characteristic upright resting position on its hammock so that its head was now pointed downward, and it was in motion as it spun silk to add to the hammock. At the same time, I noticed a *Vespula sp.* wasp (yellowjacket) flying amongst the leaves, about a meter to the left of the caterpillar. In its zig-zag flight, it happened on a straight line view of the moving larva. It was then about a half-meter away from the larva. At this point, it flew directly at the larva, grasping it at the anterior, downward-aimed end so that the wasp's head was upright. Immediately, the wasp broke off the attack and flew off, not to return. As the wasp left, the caterpillar's protruded osmeterium was exposed to my view. The wasp was in contact with the caterpillar for less than a second. The osmeterial chemicals certainly must have a rapid action! As of 9:00 a.m., on September 8th, the larva still survives, resting on its hammock."

Consequences of Environmental Resistance

Considering all of the factors that eliminate machaon individuals from the population (known collectively as "environmental resistance"), it is not surprising that each female may lay one hundred or more eggs during her adult life. Even so, due to this intense environmental resistance, machaon populations are seldom at the carrying capacity of their resources.

Ecologists determine population growth curves for various species they study. These curves take various forms, depending upon the degree of environmental resistance the population sustains, as well as on the availability of resource. A standard curve characteristic of certain basic conditions is sigmoid in shape where the vertical axis represents numbers of individuals and the horizontal axis, time elapsed (see fig. 18). Thus, the curve itself represents the rate of population size change which varies at the different phases, being very rapid where the curve line is nearly vertical at the middle of the modified "S," but slower at both ends. The carrying capacity determines the actual numbers' level at the top of the curve. This capacity level is traditionally noted as the "K" value. The symbol "r" refers to the rate of change at any stage.

Species whose populations are usually inhibited from reaching the K level are spoken of as "r adapted." Here a female's fixed reproductive resource is highly divided into many small units: ova, with minimal yolk and which may be dispersed widely, and are left unattended by their parents. This adaptation allows a rapid population increase by survivors wherever resource is freed, for instance following environmental "disasters" such as fire or flood, which wipe out much of a previous population. But most significantly, this high reproductive rate potential helps insure that at least a few individuals survive in spite of the usual intense environmental resistance, such as with the machaon complex swallowtails we have considered.[50]

By contrast, K adapted species produce comparatively few offspring, enabling them to invest much more reproductive resource in each individual. This may be the yolk in large eggs, such as in birds and reptiles, or nourishment of embryos and young, as in mammals. Here, a population is often near its carrying capacity, so that there is more intense competition among individuals in the population for the now limited environmental resource. Hence, there is adaptive pressure to best provide for individual young, enhancing their survival chances in this competition.

Let us now consider speciation and ecological niche characteristics in a K adapted group, our own genus, *Homo*.

CHAPTER NINE:
A Context for *Homo* Speciation

Before we deal with the pertinent specifics of what is known about the origin of our species, *Homo sapiens*, let us consider some basic aspects of our design and compare them to design alternatives.

One basic design aspect is symmetry. A sedentary animal, such as a sea anemone (phylum Ceolenterata) is best served by radial symmetry so that it encounters its environment from any lateral direction with equal efficiency. (Note that although a sea anemone has a mouth, it has no head.) By comparison, an animal that moves through its environment is better served by bilateral symmetry. Such symmetry allows the concentration of sense organs, etc. at a leading end.

The physical functions allowing animal life require delicate tissues and organs, such as those of the nervous system. Hardened container structures for such organs can serve to protect them. Properly shaped, hardened structures can also provide leverage for efficient, rapid locomotion. Such a skeleton can form externally (exoskeleton) as with the Arthropods, or internally (endoskeleton) as with the Echinoderms and Vertebrates. However, whereas an endoskeleton can grow from all of its surfaces in that it is surrounded by living tissues, a true exoskeleton only has living tissue at its internal surface.

Arthropods are able to increase their size only by periodic molts of the smaller exoskeleton which is replaced temporarily by a soft, elastic layer, which can be inflated and then hardened, thus becoming a larger exoskeleton which provides space for further growth of the soft, internal structures. However, there is a limit to the body size allowed by this process in that the soft phase can only support a limited bulk and weight without collapsing or even rupturing. Another limitation to size, at least in terrestrial forms, is the efficiency of the tracheal system in delivering oxygen to the deepest internal tissues.

Mollusks, such as clams and snails, also have an exoskeleton. However, it has open edges to which the underlying living tissues can add hardening deposits, allowing expanded space for body growth. (Note that especially mussels are bilaterally symmetrical even though they are sedentary in their adult phase. Apparently, their protective exoskeleton and mode of filter feeding compensate for the advantages of radial symmetry.)

In mollusks we note that, although an exoskeletal shell protects well the internal soft tissues, it can be quite cumbersome to efficient locomotion. The nautilus, a cephalopod mollusk, is able to remain suspended in open waters, and does so, at least nocturnally. It is able to do this by trapping secreted gasses in internal chambers of its shell which thereby provides buoyancy. It is able to change position by jet propulsion, forcing water from a funnel-like, respiratory organ. Thus, it can seek locations with prey and can escape its own predators. It is, indeed, bilaterally symmetrical, and possesses numerous arms, enabling capture of its prey.

Another cephalopod, the octopus, apparently has not even a vestige of an exoskeleton. This frees it for greater maneuverability. Its eight arms are equipped with suckers so that it is able to handle its prey, such as crabs. Whereas sea anemones have a mouth but no head, octopuses are mostly mouth-possessing head with attached arms. (Hence, the name: cephalo "head," pod "foot.") They have a pair of eyes structured much like our own except that the incoming light is received on the retinal surface directly rather than passing into a deeper layer. Thus, their eye design can be considered more efficient than ours. Apparently, good vision and an ability to handle objects requires a complex brain which is protected within the octopus head by cartilaginous plates. Octopuses are especially known for their learning ability and even long-term memory. Perhaps an anecdotal account will adequately illustrate this aspect.

How to Make Friends with an Octopus

Some years ago, before I retired from teaching at Western Oregon University in Monmouth, Oregon, my invertebrate zoology class and I had captured an octopus on the low tide at a rocky beach on the nearby Oregon coast. The aquarium in which we kept her had rather low sides, so that she was confined to shallow seawater. It worked out that this was of consequence. Our octopus soon learned that she could squirt a stream of water out of the tank at anyone who "displeased" her. And she seemed skilled in her aim. It seems that such behavior would be futile in an octopus's usual deep-water environment, except that it can disperse a cloud of "ink," obscuring the octopus from its predators. However, our octopus rarely used her ink, perhaps implying use judgment.

It was apparent that our octopus had learned to shoot water out of the tank at targets of her choosing which, as stated above, were usually intruding humans who annoyed her. I clearly remember an occasion when my act of removing debris from her tank provoked her displeasure. I happened to notice that she was swelling with water (which she would do to provide "ammunition"), just in time to duck away from the resulting spout. If I had not, I definitely would have been hit!

Another act our octopus learned was to "shake hands" with people. This followed her first learning to reach an arm out of the tank to accept a piece of fish or other food. When we extended a hand to her, she learned to reach an arm out to us and make contact, even without a food reward. And she remembered this behavior for the duration of her stay with us.

On one occasion, a distinguished naturalist and author of numerous books was in attendance at a conference on campus. During an intermission, I proudly introduced him to our octopus by having him reach out to shake hands. To my surprise and his alarm and considerable disappointment, she refused, and instead turned pale and, if I remember correctly, squirted him full-on! This behavior puzzled as well as embarrassed me. That night, I thought of a possible reason for it. So the next day, I had my distinguished guest try again to "make friends" with our octopus. But this time, I had him conceal his voluminous, grey beard so that the octopus would only see his eyes and upper head. I was relieved when she extended an arm in her usual, friendly manner. Apparently, a pale-grey, octopus-shape with prominent eyes is a signal of aggression to an octopus.

Note what this account indicates about octopuses. They learn and remember. They communicate, at least visually, and adjust their behavior accordingly. And it would seem that their responses, although mostly instinctive, may involve at least some degree of thought-process such as decision-making. At least we took pleasure in that possibility.

At the end of the term, I transferred our octopus to the Coast Aquarium at Newport, Oregon where she was thereafter kept in a deep-water tank.

Octopuses are likely the most "intelligent" of invertebrate animals. And their mental abilities are likely correlated with their good vision and ability to grasp and manipulate objects with their appendages. Among the vertebrates, primates are particularly noted for a combination of good vision and ability to handle objects with suitably adapted appendages, their hand-bearing arms. Although many birds certainly have good vision, they are limited in their ability to manipulate objects, using their beak or a raised, grasping foot. However, both beaks and feet in birds are primarily adapted to other functions and thus apparently not free to specialize in object manipulation. Thus, it is not surprising that primates have the most advanced intelligence.

Much fossil evidence has been accumulating that documents a rather extensive array of primate forms, now mostly extinct. It is not our purpose here to review all of this. Google searches will provide the motivated with updated references to this literature.[51]

In our context, let us first illustrate the behavior of a living primate, *Pan troglodytes*, the chimpanzee, as it demonstrates its degree of cognition and then reflect on the significance. Again, here's an anecdotal account.

Charlie the Chimp

In the mid-1990s, a colleague at Western Oregon University and I took a small group of students from our "Man and the Ecosystem" class to the Washington Park Zoo in Portland, Oregon where we had made arrangements for a guided tour of the zoo's program on endangered and threatened vertebrate animal species. This included a visit to the chimpanzee facility, part of which was an outdoor, fence-enclosed area. Our zoo guide, a woman in her thirties as I recall, had instructed us not to make provocative gestures, especially in the view of a particular male named Charlie. We encountered Charlie with his harem of several females looking at us through the fence of the enclosure. But in spite of the warning, one of our male students apparently couldn't resist making the raised-elbows, underarm, finger "scratching" signal as we stood watching Charlie and he, us. This obviously agitated Charlie. Apparently, this signal is one of aggression in Chimpanzee language. Also, Charlie had been taught sign language. (I found out several years later that one of his teachers back in the early 1970s had been Brian Johnson, an acquaintance of mine on the teaching research staff here on our campus.) At this point, Charlie signed to our guide. She interpreted his message for us. He wished her to send us strangers away and then she should come into the enclosure to join him and his troupe. As she signed her refusal to obey, she quickly instructed us all to dodge behind an obstruction. Fortunately we did so, just in time to avoid being spattered with chimp feces which Charlie began hurling at us with good aim. Apparently, he had a hoarded supply of ammunition, so we all promptly left the scene. And this is at least part of what Charlie had requested of the zookeeper.[52]

So, did the chimpanzee outsmart the human? And, if so, why was he on the inside of the cage and we on the outside? Perhaps an analysis of this question will serve to clarify at least certain principles in primate evolution.

Homo Sapiens Versus Other Primates, Especially Chimpanzees

First, let us reflect on the contrasts between us, as humans, and chimpanzees, especially as our anecdotal example illustrates. Humans have relatively larger feet than most primates. Feet that, except in rare individuals, have lost grasping ability and so are fully adapted to bipedal locomotion. (Standing on two feet obviously also requires a good sense of balance, especially when each foot only forms an effective pedestal in one direction.) Our forelimbs retain hands with the ability not only to grasp tree limbs, but also to manipulate objects such as tools. Thus, our good vision and ability to carefully manipulate objects apparently provide the potential for increased mental complexity. And, I would suggest, the human opposable thumb with its wide thumbnail is a key factor in this process.

Of our nails, our thumbnail especially, can be used to scrape and to press objects such as small fruits and shelled organisms. It seems that mental control of such actions would provide the threshold for adaptation to tool use. And extensive tool use must have been the breakthrough adaptation that has separated our species from all others on this planet. For, as we shall elaborate later, it has allowed our species to expand our ecological niche to eclipse the resources of most of our competitors.

Apparently, DNA comparison indicates that, of living primates, chimpanzees are the most closely related to us humans. In keeping with the suggestion that hand adaptation for object manipulation (in conjunction with good vision) is basic to the fostering of intelligence, it is in order that we make a comparison of chimpanzee and human hands. In nature, chimpanzees spend much time brachiating in trees in their African jungle habitat. Their comparatively long arms and short legs adapt them well to this mode of locomotion. When on the ground, they generally knuckle-walk, so that they are not well-adapted to bipedal locomotion. Also, their feet are equipped with four finger-like, grasping toes and an opposable "thumb" toe, apparently useful in tree locomotion. These foot characteristics are well illustrated in Jane Goodall's book, *In the Shadow of Man*, [revised edition, the fifth page of unnumbered illustrations titled, "Chimpanzees make nests to sleep in,"(Boston: Houghton Mifflin Co., 1988.)] In the figure opposite page 171 in the same reference, a chimpanzee is shown using a "grass tool" to catch termites. Here note the rather clumsy hands, at least compared to ours. Note also, on the chimp's knuckles, the thickened skin, similar to that on the soles of our feet. Thus, it appears that knuckle-walking involves adaptations that are in "competition" to that of tool use.

By the way, Brian Johnson tells me that Jane Goodall had occasion to observe Charlie at the Washington Park Zoo, and was impressed by his level of intelligence, good looks, and charisma relative to chimpanzees as a group. Sadly, Charlie died on September 17, 2009, apparently at near forty years of age.

Again, there is much paleontological evidence of the existence of other primate species, including other hominids that would have been much more like our species, than is the chimpanzee. Our recognition of any of these as distinct species, of course, is limited to the morphospecies concept "paleospecies." Thus, it is not in order to attempt to illustrate stage three of our speciation model here. Perhaps in time, if enough DNA evidence can be extracted from fossils, a reasonable theory may emerge. Even so, we can make observations on possible niche distinctions and the likelihood of ecological competition from what evidence is available. But first let us set the context by reviewing the current understanding of the fossil record.

CHAPTER TEN:
Speciation and Niche Distinction
in *Homo Sapiens*

Psalm 139: 14-16: I praise you because I am fearfully and wonderfully made; your works are wonderful, I know that full well. My frame was not hidden from you when I was made in the secret place. When I was woven together in the depths of the Earth, your eyes saw my unformed body. All the days ordained for me were written in your book before one of them came to be. (Bible, New International Version)

Jonah 1:17. But the Lord provided a great fish to swallow Jonah, and Jonah was inside the fish three days and three nights. (Ibid.)

We have earlier referred to what we termed the "Jonah principle." Disobedience to God earns banishment, subsequent repentance, and obedience restores favored status (for us, made possible by Jesus's sacrifice). In the account of Jonah's example of this principle, "banishment" was to a "great fish" prepared by God for the occasion. The procedure for preparing the fish is not given.

Further, we have postulated that Adam and Eve were banished from paradise (in the heaven brane) upon their disobedience and that this involved a recipient, *Homo sapiens*, here in the earth brane, likewise "prepared" by God. We now come to an account of what science has discovered regarding this preparation procedure. As we shall see, this involves "K" adaptation with eventual unique modification.

Evidence of chronological modification patterns and speciation in the primate genus *Homo* are provided by fossils and recently by DNA comparisons. In our context, to illustrate the principles involved, perhaps we needn't be concerned with comprehensive detail for the genus as a whole, but let us rather concentrate on what is known of the process specific to our species, *H. sapiens*. The following references are helpful in providing a context:

I have found an Internet posting by Dennis O'Neil, "Early Modern *Homo sapiens*," especially informative. O'Neil apparently frequently updates his Internet posting to accommodate new information. A fourteen-page posting by David Kreger provides a summary account of known fossil finds as well as an in-depth discussion of the implications. I refer here to the posting, "*Homo sapiens*" under "Archaeology, Info." Another posting, "African Eve, Eurasian Adam, the Age and Origin of the Human Species," is a five-page document by Ronald A. Fonda who mostly champions one perspective. The reader who wishes to pursue these matters in greater depth may find these references useful at least as an introduction. Here, let us overview the current status of the speciation theory, suggest the significance of various aspects, and perhaps add some speculation.

According to Dennis O'Neil, referenced above, there have been two main and competing interpretations of the available evidence: the replacement or "out of Africa" model and the regional continuity (or multiregional evolution) model. Further, O'Neil states that there is emerging a new model, the assimilation or partial replacement model. This model takes a middle ground by accommodating aspects of both of the previous models.

All of these models recognize particular aspects, such as the existence of an ancestral species, *Homo erectus*, which once was wide-spread throughout Africa and into Asia and Europe. In time, this species is somehow replaced by *H. heidelbergensis* eventually throughout its range including

Africa, Asia, and Europe. Later, an African derivative of heidelbergensis, *H. rhodesiensis*, occurs. Eventually, this new species expanded its range into Europe where it intermixed reproductively with extant heidlebergensis. The resulting hybrid mix became a hybrid swarm whereby the parent species lost their distinctions. Apparently, this stock eventually produced the species, *H. neander-thalensis*, which spread throughout Europe and the Middle East.

Meanwhile, back in Africa, a population of rhodesiensis became isolated and bottlenecked so that all the male lines but one, known as "Y-chromosome Adam," were eliminated. Just where in Africa this took place is not presently clear. Perhaps it was the speciation event of *H. sapiens*.

In time, the African continent became arid so that what had been fertile plains and perhaps jungle was replaced by desert. According to Curtis W. Marean in his article in the August, 2010 issue of Scientific American, "When the Sea Saved Humanity," this desertification of Africa began about 195,000 years ago during what has been termed the "Marine Isotope Stage 6" (MIS6). This was a long glacial stage that apparently lasted some 72,000 years. According to postulations by climatologists, this period saw desertification significantly expanded compared to the African deserts of the present interglacial period.

Apparently, both rhodesiensis and early sapiens mostly depended upon large animals as prey and must have also gathered plant foods. But these resources must have greatly diminished with the desertification. However, there is good evidence that at least one remnant population of sapiens persisted, this near the southern coast of Africa in the vicinity of Pinnacle Point in the Cape Floral Region. Marean states that the population of *H. sapiens* probably dropped from some 10,000 breeding individuals prior to desertification, to "just hundreds." He has studied fossil and artifact evidence at this site which suggests that here there was access to seafood resources and a rich supply of edible plants, particularly "fynbos" and "renosterveld," shrubby plants that form edible geophytes: bulbs and corms, these nutrient-rich in carbohydrates. However, large prey animals would have been absent. And hence there would have been adaptive advantage for a shift in eco-niche for this population. This must have involved new technology which would likely require shifts in social structuring, all dependent upon genetic change.

One would expect more frequent inbreeding and perhaps the concomitant fixation of alleles in this smaller population which may have enhanced the adaptive shift to the new niche. During periods when the seashore receded, the inhabitants of the caves at Pinnacle Point would have to know when to travel the distance to the shore to take advantage of low tides to access intertidal food organisms such as shellfish. This would require awareness of the tidal patterns, ability to keep records, and to communicate the information to others. Apparently, fishing technology was eventually developed to enable access to offshore food resources. This may well have involved some sort of boat building. And, of course, success in this regard would enable exploration of new areas.

All of this would certainly have involved social structuring conducive to cooperation even with non-kin. Different individuals or groups would likely specialize in certain aspects of the technology and its use. By cooperating with other groups specialized in other aspects, all would benefit. And what of tool use and its enabling of technology (to be considered shortly)?

As the glaciers retreated, so would the desert, so that the remnant humans could then expand their range and numbers. This expansion eventually involved invasion of Europe and Asia where Neanderthals still lived.

Early Stages of *Homo Sapiens*

It is well-established from archeological investigations that especially early humans used tools made of stone and of bone. Later, iron became an important tool-making resource, apparently first acquired from iron-rich meteorites. And the use of fire, which could be started by sparks from the manipulation of certain stones, was certainly an important step in the development of human culture. Eventually, fire was used to refine the production of iron tools.[53]

With the advent of extensive tool use, learning to produce and use various types would foster specialization by individuals within the community. As stated above, this in turn, would provide adaptive advantage to social cooperation, even outside of close family lines. And it seems obvious that the most efficient social groups, usually called tribes, would have advantage in competition for resource, probably usually involving control of expanded territory.

It is my understanding from various accounts that primitive human tribes, such as have been studied in New Guinea, compete for territory and resources. Eventually, a few, then one, becomes dominant to the exclusion of the rest, expanding its territory extensively. But with limited transportation technology, resource is best used locally. To adjust, the winning tribe soon subdivides, usually along family lines. And the pattern repeats. Thus, the genetic base for increased social efficiency, including the use of tools, would have selective advantage. And all of this would certainly foster increased intellectual ability. Hence, *Homo sapiens* apparently has changed from its original characterization at its speciation event. At what point would our species be adequately "prepared" for the reception of the essence of disobedient Adam and Eve in keeping with our Jonah principle? Does our species' unique practice of religion have a role? Before we deal further with such questions, let us first elaborate briefly on our species' interaction with other hominids.

Of particular significance here is the interaction of humans and Neanderthals as populations of our species spread into Europe and Asia. Apparently, there is little evidence, from comparison of the genome of modern humans with the recently revealed genome of Neanderthals determined from fossil remains, that there was any extensive interbreeding between the species. What may have constituted the reproductive barrier? It may not have been entirely physical, in that there apparently is at least some evidence of gene exchange. Perhaps it was primarily social. (One is tempted to speculate on the potential for the success of a movie based on the idea of such interaction between populations of the two species.) There would have been overlap in the niches of the two species, humans certainly making use of larger game now, as well as of seafood. But humans likely won the competition, primarily because of their intellectual superiority and the technology it fostered. However, as we shall elaborate later, such technological ability has the potential for serious consequences.

At least one other species in our genus was once contemporaneous with our species. I refer here to *Homo floresiensis*, fossils of which have been found in a cave on the island of Flores in eastern Indonesia. However, according to various Googled postings on the Internet, there is some controversy as to the legitimacy of these fossils representing a distinct species, some holding that they rather represent pygmy-like *H. sapiens*. One posting holds that, in fact, there is a nearby village (Rampasasa in the Weamulu region, Waerini subdistrict) in which there presently lives a group of native pygmies, definitely of our species, but somewhat larger than the single, well-preserved female skeleton so far found in the cave. And apparently, further investigation in the area for fossils and artifacts has been disallowed so that further research has been impaired. However, the one nearly complete skeleton has been determined to have a mix of *Homo erectus/H. sapiens* cranial features with some Australopithicus skeletal features. And it appears that the species died out before our species reached the island, so that any interaction of the two would not have been possible. One hopes that further investigation of even other nearby islands will help clarify the situation.

The Point of Cognitive Sentience

It is my opinion that the origin of religion indicates the point at which our species became at least sufficiently sentient to serve as an actualization of God's eternal intention of willfully obedient beings. And, perhaps, at this point, our species would be fully "prepared" for the reception of fallen Adam and Eve's essence ("souls"). Although Marean's study of the Pinnacle Point cave artifacts indicates that the ancient inhabitants had a sense of symbolism evidenced particularly by their use of primitive paint, he does not cite evidence of religious practice. Nevertheless, one suspects that some form of religion may have occurred here in the Cape Floral Region along Africa's coast during MIS6. From a biblical perspective, this is of significance if, indeed, all of the present, widely distributed populations of humans originated here. Let us now consider why our species is unique in the practice of religion. There are several factors involved. First, the basic condition.

According to Newberg and Waldman in *How God Changes Your Brain*, 2009 (New York: Ballantine Books 43-44, 2010 edition) an apparently unique aspect of our species which may account for the universal practice of religion, at least among primitive societies, is the adaptive condition of several parts of our brain. For example, the anterior cingulate allows us to experience God as loving and compassionate.

Wherein is this Adaptive? Human Society as "Socionic," the Basis for Civilization

Some years ago, a TV program known as the *The Six Million Dollar Man* was popular. In this fictional setting, the hero, played by Lee Majors, had been given mechanical supplements for various damaged body parts. This purportedly allowed him superhuman powers, and hence the drama of the show. May I suggest that the same principle can be applied to human social structure, and this in real life. Mechanical devices, such as tools, and the use of energy to operate them, constitute a supplement to human social structure giving human society a dimension which is unique to our species. Hence, the term "socionic" in characterizing the human social system. I argue that this is a foundational principle in our species' success in ecological competition with other species. And, I argue, it entails a situation which provides adaptive advantage to the practice of religion.

CHAPTER ELEVEN:
Socionics and the Tragedy of the Commons Principle

In 1968, Garrett Hardin published an article in *Science* (162 (1968):1243-1248), titled "The Tragedy of the Commons." To illustrate the principle involved, Hardin uses the example of a pasture open to all herdsmen in the community. Each herdsman realizes that the more cattle he can put on the pasture, the more profit he can earn. And the system rewards the greediest of the herdsmen, in that they have more cattle to sell than those who realize the limits of the system and therefore, pasture fewer cattle. But as this practice eventually depletes the pasture resource, even the modest herdsmen suffer the consequences, and, in fact, end up poorer than those who misused the commons. Hence, the tragedy: those who misuse the resource are rewarded, those who use it carefully are penalized.

Now, there may be means whereby the "tragic" consequences of this process can be avoided. For instance, suppose the pasture were to be subdivided and each herdsman given the use of a private plot. Here the reckless users suffer the whole of the loss, resulting from their misuse, and the careful users sustain their modest reward. However, not all resource commons can be as easily divided into private portions. We shall suggest other means of avoiding the commons tragedy shortly. First, however, let us consider why such use of a commons can thus be tragic, whereas competition for resource generally functions positively in natural selection, allowing survival of the genetically "fittest." Why aren't the greediest the fittest in the human situation?

Territoriality in Non-human Species versus Humans

It is of pertinence here to recognize that many species of animals essentially privatize resource units by maintaining territories. Many examples could be cited such as with nesting birds or various mammals such as wolf packs. Note, however, that here each territory is "owned" by a group of kin, so that its successful maintenance benefits the genetic base of the group. By contrast, with humans, a privatized resource may be utilized by groups of mixed kinship, particularly in broader social systems. As we have seen, cooperation across kinship lines allows for the comparatively great complexity and concomitant efficiency, and therefore success, of the socionic system. However, we have also seen how the efficiency of the human socionic system has the potential to deplete the resource of a commons. Thus, such a system is apparently unique in nature and negates the normal selective advantage of "greediness." How then, is human society to avoid this potential?

Other Means of Preventing Tragedy in Commons Use

Let us return to the pasture commons example. We have noted already that such a commons could be divided up into privately used parcels rendering the resource no longer a commons. What other ways might tragedy be avoided even if the commons is maintained? Could not the users agree to limit their use to a sustainable level? But what is to ensure the participation of all individuals in the community's agreement? And how is a nonconformist to be dealt with?

One evident answer to the dilemma is to form a government that establishes laws concerning the agreed upon use of the commons. And this is obviously done extensively in human society. If offenders are caught, they must suffer appropriate punishment. And if this consequence is modelled well, perhaps its motivating effect will offset temptation to break the laws of the agreement. But in situations where apprehension is sufficiently unlikely, the system still will reward those who can violate the agreement, and thus the tragedy is not avoided.

And what of individuals who serve in a controlling role in the social system who give themselves, and perhaps their friends, special privileges? Some human social systems consider such individuals as royalty and accept the practice.[54] Others consider such action "corruption" of government. But, as we shall see, the "plot thickens."

In certain types of government, the "greed" motive is utilized to stimulate innovation, especially in the mechanized aspects of social structure, with the expected outcome of benefit to the system as a whole. (However, some entrepreneurs may actually use the potential community benefit rather than greed as their motivation in such endeavors.) Such innovation may enable the use of new resource and, in the process, expand the effect of the socionic system upon even new commons. Usually some form of control is imposed to moderate the consequences. The recent production and use of wind-generated electric power may serve as a suitable example here. Perhaps in such situations, the "greediest" are indeed the fittest, at least in social function. It would be interesting to determine the extent to which genetic advantage may result.

Wherein is Religion Adaptive?

Of course, human social systems vary from the simpler to the highly complex. More basic systems, such as characterized the original Native American cultures, were generally well-balanced with their resource supply; this perhaps largely due to the social structuring effects of their religion. Often in such systems, offenders of the "rules" are believed to be vulnerable to punishment by the all-seeing, ever-present, spirits. (Perhaps the most effective control of commons use occurs where the tenets of religion and government imposed rules are consonant.)[55]

One might argue that when more complex and expanding socionic systems take advantage over simpler, balanced ones, there is, in effect, a breach of control in the use of commons. Here, rather than competition among individuals, whole social systems compete. The encounter of the established, Native American social systems with the invasive, European, more complex socionic system serves as an example of the outcome of such competition. In the eventual result, perhaps tragedy is imminent to the extent that balance is forfeited. The present, rapidly expanding human population certainly attests to the loss of balance.

But then, socionic systems with more complex technology characteristically are able to utilize new resource that is otherwise unavailable and thus are sustained, at least for a time. Perhaps here is the point at which we should elaborate further upon the pertinence of the population growth curve to the present human population circumstance.

Stages of the Population Growth Curve and Their Significance

Figure 18 shows typical population growth curves. In these standard situations, note that a population may expand rapidly for a time, but eventually grows with decreased rate until expansion ceases. As stated earlier, this level is known as the carrying capacity and is symbolized by "K." (Thereafter, the population level decreases if resource is depleted.)

Let us here consider the perspective of individuals in the context of each of these stages. When a population's resource is underused, that is, when demand is below supply, competition among individuals is minimal. It is relatively easy to meet one's needs. And in socionic systems, there is good opportunity to exceed survival levels of resource use. That is, the "standard of living" can be high, perhaps even excessive. (Let us here introduce symbols to represent this system. Let "s" represent the composite survival needs of the individuals of a population and "e" represent the excess use of resource that allows a socionic system. Thus, $s + e = u$, where "u" is the total use of resource by the population.)

On the other hand, if demand exceeds supply, competition may become intense. Usually in non-socionic systems, dominant individuals utilize sufficient resource levels to allow them to survive and thrive at the expense of weaker individuals who consequently may not survive. The "greediest" are then indeed the fittest, resulting in the perpetuation of a healthy core population. Thus, the genes that allow such, perhaps, socially dominating traits are naturally selected to constitute the gene pool of the species.

However, as considered earlier, socionic systems encounter problems with such lack of resource in that the whole technological component of the system is undermined by the reduction in the value of e. And, in that the socionic system as a whole is dependent upon the use of "unnatural" resource, when this resource is not accessible, the artificial K level (explained below) drops and now even the s value decreases. Runaway feedback may result and the whole socionic system may collapse. However, usually such conditions are local, so that individuals may escape to more stable areas or other populations that remain functional may send "relief" resource supply. If, however, nearby populations have been in competition (are considered enemies), then the unfortunate collapsed system's individuals may perish.

In these dire circumstances, there may be motivation for dominant socionic systems to thus "outcompete" the weaker systems for the limited resource. Hence, the most socionically fit systems survive at the expense of the less-effective systems (including the individuals of which they are composed). And so the same principle applies, but here at the population unit level, rather than the individual as in non-socionic systems. The most effective socionic system "survives and thrives" and perhaps the genes that allow such dominance are thus naturally selected to constitute the gene pool. This scenario was perhaps the most likely outcome in primitive human social systems as was suggested earlier.

Inflation of the K Value

On the other hand, as implied above, the socionic nature of human social systems allows another solution to the resource shortage dilemma. Advancing technology may allow the further inflation of the K value by access to new (unnatural) resource. Thus, what would have been the peak phase (K) now falls in the pre-peak zone where there is still good supply of resource. This allows the perpetuation of high e values. However, there is also the likely increase in population level, so that the increased s demand competes with the e value and there results a runaway feedback system dependent upon artificially increasing the value of K. Note that whenever $u = K$, the growth curve peaks out and will thereafter decrease, unless K can be elevated (or u stabilized) because u cannot exceed K in a stable system. Of course, runaway systems must eventually collapse in that all supporting resources are finite, a tragic outcome indeed.

Behavior: Instinctive Versus Learned

Another important aspect of the human socionic system is the repression of many instinctive behaviors and their replacement with enhanced ability to learn. For instance, whereas in most species non-kin males compete, in humans such instinctive behavior can be replaced with learned cooperation.

Instinctive behavior generally involves response to stimuli known as "releasers." For example, male butterflies of many species have a pair of scent patches (androconia) on their wings. In courtship, contact of the female's antennal knobs with these patches allows their chemical content to stimulate or "release" the next behavioral phase in the courtship process. In birds, visual stimuli such as the display of color patterns of the male's feathers, act as releasers for subsequent female response.

In humans, female body shape, voice quality, texture of hair and skin, and perhaps pheromones may act as releasers for male response. In males, facial hair, voice quality, body shape, etc. perhaps may serve as counterpart releasers. In most animal species, such behavioral patterns are automatic, that is "instinctive." They are genetically based and hence adaptive. In humans, response may be modified by mental control and eventual learned, altered (or "conditioned") response. But any genetic alterations that would eliminate all response motivation would be lost from the gene pool (unless by recessive alleles). So how is it that, for instance, a human female can be "privatized" in a socially interactive community? And wherein is there "tragedy" if a female (or male) were, indeed, a commons?

As has been stated earlier, humans are highly K adapted in their reproduction. Relatively few offspring are produced and require high levels of care over extended periods. If both parents participate in providing care and resource, the reproductive success is enhanced.

Reproduction from the Perspective of Reproductive Behavior Gene Perpetuation

Any care a mother provides her own children obviously enhances the perpetuation of her own genes. But, if the care-providing male is not the actual father, then his genes are not directly benefitted from his efforts. On the other hand, if the female can be removed from access to other males, so that he is the actual father of all of the female's young, then his genes for nurturing effort survive, along with the mother's, and the results of both of their nurturing efforts are maximized. However, the situation is complicated by the fact that if he can leave genes by fathering children by other women, his reproductive behavior genes become even more prominent in the gene pool. But then, this creates a problem for the men nurturing the young who are not their own as stated above. And some men could leave their genes for behavior by fathering children with several women, making no contribution to the nurturing of their offspring, leaving that to each mother and her male provider. In other words, they would be "parasitic" in their male role. But if such a male was sensed by females as of superior quality so that a female's offspring would have his superior traits, then it would be to her advantage to mate with him. But, again, her male provider's genes would be disadvantaged.

Apparently, there have been examples of human cultures in which a woman's brother provides male nurturing to her offspring who would share some of his genes, even though he is not their father. In turn, he leaves his genes directly by mating with females in other such families. But then, his nurturing efforts support fewer of his genes than if he spent them on his actual offspring.

In situations where there are more females in the adult population than there are males, such as when many males are lost in warfare, then polygamy may have an advantage. More females would have opportunity to leave their genes and get at least some help in the nurturing of their offspring from the group's male and perhaps from the other females. Polyandry would certainly enhance the degree of nurturing for the offspring of the group's female, but puts the males at disadvantage in that they each could father fewer children.

Perhaps, then, it is not surprising that most human social systems have rules that allow the privatization of each female and male so that they are not a commons in the system. These "rules" are generally supported by both the system's government and by the "morals" promoted by the religious components. Rules for male behavior are, perhaps, less easily enforced than for females, but if broken, would, of course, disrupt the privatization of females in the system. Let us note here that certain other animals, such as penguins, also have privatization behavior in mating pairs, even though they lack socionic social structure.

In regards to the extent of nurturing allowed by the one father, one mother family, it is of note that in the human socionic system such nurturing may extend well beyond the female's reproductive stage. Economic and other help, such as grandchild care, can be extended to the couple's offspring by enhancing their socio-economic positioning in the system and thereby the perpetuation of their shared genes. This help can be of important benefit, especially if competition is intense. Note that it gives selective advantage to family stability even beyond the years of direct offspring care. (It is interesting that the principle is expressed even in plants such as strawberries, where a parent plant extends "runners" that root in adjacent space, even where other plants are already established. Nourishment from the parent plant gives the "offspring," here a clone, competitive advantage in the tangle.)

Is Homosexuality Adaptive?

If reproductive behavior in humans is genetically based as in other species, what might be the adaptive significance of homosexuality? Judging from Google references (mid-2011), apparently no chromosome locus has yet been demonstrated for a sexual preference gene in our species. However, if one is eventually discovered, it is tempting to predict that it may be of similar adaptive function to that of the gene for sickle cell anemia.

In sickle cell anemia of humans, apparently two co-dominant alleles determine the shape of erythrocytes (red blood cells). If the normal allele is homozygous, then the erythrocytes are of normal shape and function, but the individual is vulnerable to malaria and its often deadly consequences. On the other hand, if the sickle cell allele is homozygous, then all of the erythrocytes are fully sickle-shaped, causing them to have failing function. Such individuals generally die at a young age without reproducing. However, individuals who are heterozygous for the alleles produce near normal-shaped, functional erythrocytes that become sickle-shaped only if infected by the malaria parasite, *Plasmodium* (four species including *vivax*). These altered cells then cause the contained parasite to die, which controls the infection, allowing the person to survive and successfully reproduce. Thus, it is this genotype that perpetuates both alleles, at least in geographic areas where malaria is prevalent.[56]

Suppose now that there is a similar genetic base for sexual preference in humans. Individuals homozygous for the same-sex allele would leave no offspring. Individuals homozygous for the opposite-sex preference allele would have offspring, but, especially the males, would tend to compete rather than cooperate with individuals of their own sex, thus reducing the efficiency of socionics. But heterozygous individuals would both successfully reproduce and cooperate closely with others of their own sex. This should foster well-functioning socionics and thus perpetuate both alleles. However, there are obviously compensating factors which foster cooperative social interaction among even males homozygous for heterosexuality. For instance, covering body parts that act as sexual behavior releasers with appropriate clothing would serve this function. Perhaps even the predominant practice of preventing beards in men by shaving serves this role in part.

Thus, it would seem that any gene for homosexuality would have limited selective advantage, at least in modern human populations.

Adoption and Institutional Child Rearing

If the adopted or institutionalized children are of appropriate genotypes, then these practices should not adversely affect a gene pool. Presumably, this is most often the case so that gene pool deviation effects are minimal.

CHAPTER TWELVE:
Social Symbiosis or "Sociobiotics"

Let us next briefly consider the role of other species in the success of human social systems. We will begin with plants, again concentrating on principles rather than extended details.

Plants were important to primitive humans as a food source, but also in the production of tools and shelters, and as an energy source providing fuel for controlled fire. And, of course, we continue our dependence upon plants in similar ways. Now our advanced technology allows us to farm the species we find useful, usually to the exclusion of species we consider un-useful. But this exclusion of some species, if extensive, leads to a problem: the potential extinction of these species. (Of course, this principle holds for animals as well.)

Let me include here an anecdote to illustrate this problem. Recently, I was a participant in a meeting involving the likely effects of listing a small, blue butterfly as endangered. The potential land use proposed for its unique habitat was cattle grazing which likely would require converting the habitat into grassland. Those of us proposing the listing were convinced that such use would result in the elimination of the butterfly's niche, which would certainly lead to the insect's extinction, in that the known range of the species is limited to the area involved. During the meeting, a young member of the group advocating grazing the land raised the question, "Why should it matter if another species goes extinct? Is not eventual extinction a normal process in nature?" Implying that, after all, why should a small butterfly's needs come before the economic gain that would be provided by the proposed grazing. And everyone in the large conference room looked at me for a profound answer. Earlier in the meeting, the group proposing the grazing use had emphasized their gratefulness for the provisions in nature granted by the Creator. My reflex response to the questions proposed was "You will have to take the matter up with the Creator." This resulted in silence, and the issue was dropped.

But there is more to the matter, of course. One certainly can argue, from the perspective that we have suggested earlier in this treatise, that the existence of any species would be an actualization of God's intention, and who are we to challenge his wisdom? (Sometimes this may be hard to understand, for instance: God, why mosquitos?) But might even butterflies have a significance even to our species' well-being?

Biologists understand that ecosystems are complex, involving many species of plants and animals. And some ecosystems may have unique aspects, which may well be the case in the pumice soil-based system here to the east of Crater Lake, Oregon. It was a hobbyist who discovered this little butterfly in his recreational pursuit. And knowing now that there is one unique component to the ecosystem, could there not be others? Perhaps even species of potentially important use in the future of our species' advancing socionics? Is it not wise, then, to preserve and protect even a small butterfly (and its habitat) that serves as an indicator species for what may well be an ecosystem rich in potential biological resources? For instance, there are many species of native bees found here. Perhaps some have important potential as pollinators of future crops. And what of the potential usefulness of unique DNA in future genetic engineering?

The Significance of Preventing the Extinction of a Plant: Meadowfoam

The following is an example of the near loss by extinction of what is becoming an important plant resource:

There are several species of meadowfoam of the genus *Limnanthes* (family Limnanthaceae) known from western North America. Some, such as *L. gracilis* are very rare, and *L. floccosa californica* is apparently now listed as endangered. Suppose these plants were to become extinct. Of what consequence would that be to us humans?

The seeds of another species in this genus, *L. alba*, have recently been developed as a source of high-grade oil, in part replacing the once used, high-quality oils derived from certain whales. Some farmers here in the Willamette Valley of Oregon have rather recently begun to farm this species commercially. When in full bloom, the fields are snowy white, certainly giving the fields a "foamy" appearance. But there is apparently still need for genetic improvement for the desired oil qualities. And this is where the other species in the genus may play a role. Presumably, by controlled cross-breeding, their gene pools may be accessed, which may provide the desired qualities. Obviously, the extinction of any of the species would reduce the options.

To set seed, the vast numbers of flowers that make up the foamy meadows require cross-pollination. Might it be that some species of native bee could someday provide the service more reliably than the introduced honey bee now used? Thus, we see illustrated the potential practical role, in addition to the ethics involved, in protecting endangered species and their ecosystems.

Let us now consider some principles involved in the role of animals in human biotics.

Sociobiotics: The Role of Animals

Primitive humans obviously used other animal species as a source of food, but also as a source of clothing and tools. Hides properly treated could not only be tailored into clothing, but could even be formed into shelters. And bone products could be used as tools such as spear tips and for digging roots, etc. In that clothing and shelters played a role in microclimate control for early humans, it would, at least in part, allow the expansion of the range for our species, even into regions with colder climates. And the domestication of certain animal species such as cattle, horses, and dogs, would enhance the effectiveness of the social systems of our species by providing food products, transportation enhancement, and increased sensitivity to the presence of prey and predators, the latter in that dogs have a more keen sense of smell and hearing than do humans.

In such systems, the involved animals are controlled by the humans who claim their ownership. They are bred to better serve the function of the human social system upon which they thus become dependent. One could say that rather than fitting a natural eco-niche, they are force-adapted to a "sociobiotic niche," its parameters determined by another species, *Homo sapiens*.

Let us note here that, to some degree, the principle of domestication is invoked by some species other than humans. For instance, certain ants "domesticate" aphids and even certain butterfly larvae, e.g., some Lycaenids. In these cases, the ants use secreted products ("honey dew") provided by their subjects, and in turn, protect them from parasites and predators and may even transport them to good food sources. And certain ant species may "enslave" members of other ant species, which are used to perform duties that enhance the social system of the captor species. Again, it is the principle that is of pertinence in our present context. The fascinating details of such systems can be researched elsewhere. (For instance, E. O. Wilson has published extensively on these matters, especially the social systems of ants.)

One can speculate that our species may have once enslaved other species in our genus such as *Homo neanderthalensis*. But more recently, our species has certainly practiced conspecific slavery to varying degree. If one defines slavery as "induced performance of a subject for the benefit of the inducer," we can apply the principle perhaps more broadly than is traditional. Usually, we think of slavery as forced performance of duties by those enslaved and whose food and dwelling needs are provided by the owners. Essentially, the slaves fill the role of domesticated animals.

Apparently, in primitive human societies, one tribe treated other tribes as functionally non-human. Tribal body markings, etc. were used as indicators of who was "human" from each tribal perspective. Thus, capturing and enslaving individuals of another tribe could be considered to be in the category of "domesticating animals." In some cases, the practice apparently went well beyond forced performance of duties so that such "domesticated animals" could even be used as food. This, of course, is what is called cannibalism.[57] Apparently, it sometimes also involved treating those of other tribes as prey to be hunted. And this was sometimes done in creative ways as described in the book, *Peace Child*, by Don Richardson (Ventura: Regal Books, 1974).

Again, if one defines slavery to include any induced rather than only forced performance, then the modern practice of hiring workers could be considered a form of slavery. [58]However, rather than concern for the feeding and housing of one's employees (part-time slaves), an "owner" simply pays them a wage, requiring them to fend for themselves when on their own time. And thus, we see one reason that the practice of using monetary systems is of great significance in the function of modern human social systems.

The Significance of Social Positioning

Earlier, we considered the need for government with its leadership in the structuring and function of human social systems. It is apparent that such involvement is vulnerable to the inequitable limiting of rights and privileges, especially to members of lower standing in the system. Further, in the more primitive tribal social systems, family line affinity apparently played a major role in group identity. However, in more complex social systems where family lines are socially mixed, other means of group distinction develop. Especially, the socionic nature of human social structure may give opportunity for non-relative, subgroup formation such as along functional roles in the system.[59] For instance, farmers may relate to each other differently than they do to factory workers. Other means of subgroup formation would include factors that determine perspective, such as political party affinity, and especially religious affiliation or lack thereof. And it may well be that religious affinity has played the most significant role in human history.

To the extent that resources are a commons, when they are in short supply ("u" is nearly equal to "k," see chapter eleven), and therefore competition is intense, then social group interaction is likewise intensified. For example, who should reduce their standard of living ("e" value) more, to compensate such as by paying higher taxes, to maintain the socionic structural base which, if lost, would cause the collapse of the whole system and the likely demise of its members?

Other factors in subgroup interaction involve each group's efforts to expand their membership and influence. And this may especially involve means to prevent the loss of members. Perhaps such interactions are most intense among religious subgroups. An individual's rights within a group may vary concomitant with their degree of submission. Human history is rich with examples. The following phrases are indicative of such treatment: "Help get me off the hook!" "Give him the third degree!" "Meet the iron maiden!" Some "non-conformists" at one time were even burned at the stake. But even the practice of shunning can be effective in preventing further loss of a group's membership.

Note that all of this manner of treatment is in stark contrast to Christ's teaching and modeling. He who commanded that we love our neighbors (even those outside our subgroup) as we love ourselves. He who taught and modeled that a leader be servant to the group, not its exploiter.

Freedom of Religion versus Freedom from Religion

In reflecting upon the basic tenets of the society in which I live here in the United States, I conclude that freedom of religion is of basic significance. In such a system, one expects that a religious group's viability would be determined by its "attractiveness" as compared to forced compliance. And it would seem that "freedom from religion" would, in time, be extremely disruptive. On what basis would I "do to my neighbor as I would have him do unto me"? Why not, rather, take advantage of him if it would better my socioeconomic position? Would not the greediest thus become the "fittest" and thereby threaten the socionic structure?[60]

Further Socionics Summarized

The extended details of the involvement of technology in human social structure, here termed "socionics," of course, are treated extensively elsewhere, but in works of thesis other than the present. Here let us give a brief summary of the significance from the socionic perspective.

Primitive human social systems began to use tools for hunting, gathering, and preparing food. Clothing was probably first used to especially hide male genitalia so that cooperation in social function was enhanced. Again, here's an anecdote to illustrate.

Years ago, I heard an account of two primitive tribes in New Guinea. Christian missionaries, in outreach to one tribe, had suggested that clothing standards were important in promoting "modesty," expecting the tribe to adopt theirs. However, in response to the missionaries, the tribe's men, who traditionally wore only penis sheaths, expressed sincere concern for the men of the other tribe who wore nothing at all. And they apparently considered ways to get these naked men to realize that they were immodest and should also use penis sheaths. It would have been most interesting to compare the degree of social cooperation within each tribe in the initial condition, but apparently this opportunity was overlooked. I forget the outcome of the account, but I assume that the missionaries eventually convinced the tribesmen of both groups to adopt more advanced modesty. If so, one would predict that social cooperation was thereby promoted.[61]

Some issues of *National Geographic* have shown women's clothing standards in primitive cultures. Often women in the pictures wore only skirts, leaving the upper half of their bodies exposed. Presumably, this was sufficient to allow some degree of privatization in the social structure.

But clothing can have other functions, group, and even social role identity included. The use of uniforms for group identity, even in modern society, exemplifies the principle. But perhaps that with the most potential, from a socionic perspective, would be the microclimate effect provided. Primitive cultures usually use various types of shelters, initially as protection from carnivores and scavengers and the intrusion of competing humans of other groups. But, as mentioned previously, both personal clothing and the building of shelters have the potential to protect from inclement environmental conditions that exceed the microclimate tolerance of the naked human body. And thus, the elaboration of clothing and sheltering structures would allow the extension of our species' range into other geographic areas, and now even into space. For even mobile structures, such as automobiles and space craft, are an extension of this phenomenon. And all are aspects of socionics which require cooperative roles in social structure, even beyond family lines.

Transportation in appropriate devices, of course, not only serves the comfort of travelling humans, but also the delivery of materials and goods as needed to support socionic structure and function. We must also recognize the important role of enhanced communication in the function of socionics.

And all of these functions in some way depend upon energy expenditure. Thus, restriction of energy supply can have dire consequences. One speaks of a system's "energy base" with real significance. Of any factor involved in advanced socionics, the loss of its energy base would certainly threaten the whole system. If "u" diminishes to the extent that resource support for "e" is lost, in that "e" allows the existence and function of the socionic structure upon which the population depends, the population level "s" would diminish as well. (See stages of the population/socionic growth curve, chapter eleven.) This would likely result in runaway feedback so that the whole system would collapse, unless there were intervention from the outside. Obviously, strong dependence upon non-renewable energy sources does not bode well for the future of socionic systems. Although temporary availability has allowed the establishment of our system, there must be compensation for its eventual exhaustion if socionics is to persist here on Earth.

CHAPTER THIRTEEN:
"Thy Kingdom Come, Thy Will Be Done on Earth As It Is in Heaven"

Wherein is God's kingdom to come to Earth? The Bible is clear on the matter of God's involvement in nature such as watching sparrows and sending rainstorms. Psalm 145:16 states that the Lord satisfies the desires of every living thing which must include even protozoa. So isn't the whole of the universe, including planet Earth, already created as God's natural kingdom? It seems, then, that the request here quoted refers to the willful obedience of us sentient beings to God's will so that his intentions for Adam's descendants are actualized. Sentient beings endowed with free will choosing to honor God in obedience, thus becoming accurate images of that part of his sentience that we have been created to reflect. Logically, God cannot tolerate false imaging if his integrity as actualizer of his will is to stand. This obedience, including even our social interaction, thereby determines even the structure of human society.

The above chapter title quote from the Lord's Prayer (Matt. 6:10) certainly indicates that there is the potential for human society to become as God wills it. But what is God's will for us humans on this planet? How is it to be realized? How can we use our free will to obey if we do not know God's will for us?

According to Genesis, Adam and Eve disobeyed God's warning and were therefore expelled from the Garden of Eden and condemned to "death." Our present thesis holds that the Garden of Eden was (and is) in the universe brane (the heaven brane) that is opposed to the earth brane that contains our planet (using the ekpyrotic universe model of Steinhardt and Turok). Adam and Eve consequently find themselves on this planet, either in essence "breathed" into a certain pair of *Homo sapiens* (soul transfer, see chapter fourteen), or perhaps as sample members of our species, who were first placed in the Garden of the heaven brane. Then, after the fall, they were returned to planet Earth to bear children who interbred with the earthly members of our species whom God had prepared (created) for this purpose. Thus, sentience would have originated in *H. sapiens*. But we are naturally destined to die, physically and "spiritually." As fallen descendants of Adam, we presently must repent and obey to be reinstated as accurate images of God, such compensatory option having been provided by the atonement sacrifice of God, the Son, who was incarnated into our species as Jesus Christ. Thereby the death to which Adam and Eve were condemned has been conquered. Thus, God's eternal will that we, as free-willed beings, obey him and thus reflect his true image can yet be actualized.

And how then is God's will to be revealed to us humans? In the biblical theist community, it is generally recognized that certain individuals, having been inspired by God, have expressed these inspired thoughts in writing, eventually compiled as the Bible and thus, recognized as the "Word of God."

A prominent theme in such messages, having been recorded over the ages, is the significance of an aspect of God coming to our level of existence as the (incarnate) Messiah. The New Testament of the Bible proclaims Jesus Christ to be the Messiah. And thus Christ's message to humanity is an expression of God's will, this message recorded for us by the New Testament authors who were contemporaries and included actual witnesses.

God's Will Revealed in Specifics

We see, then, that God's will, in principle, is revealed in the Bible. But wherein is it revealed in specifics? Should I mow my lawn today? Which needy neighbor should I help and how? I suggest three basic aspects to the acknowledgement and obedience of God's specific will that are pertinent here. First, an intellectual (mindful) application of biblical principles to the circumstance. Second, a motivational feeling promoting a loving response, perhaps best sensed through prayer. Third, an unfolding of opportunity-providing circumstances.

The Bible states that God "clothes the lilies with glory" (Matthew 6:28-30, Luke 12:27-28). Is it within his will, then, for me to mow those in my lawn? (One might argue that lilies have a proper place for thriving which isn't in my lawn. But what of actions that may lead to a species' extinction?)

Suppose I have a neighbor who is about to become homeless due to financial problems. Should I share my finances with him or her? But what if this neighbor is lazy and won't work to earn a living? Or suppose they want to work, but cannot find employment? Christ specifically emphasized the biblical commandment to "love our neighbor as we love ourselves." How, then, does this principle apply here? Is it "loving" to enable a neighbor to be slovenly? How might "tough love" apply? Should such a person simply be abandoned and avoided until they accept responsibility? And what if the needy person wants to work but is "unemployable" due to physical handicap? And what if the handicap is due to bad choices, such as drug use or even over-eating? Here may I use an anecdote to illustrate.

Recently, a person who is part of our church subgroup announced in class that she was about to be evicted from her home due to financial need. She emphasized that she had applied for employment repeatedly, but had not been accepted for any job. It happens that she was overly obese, which obviously limited her physical abilities. What should we, as group members, do? Some held that this person should not be enabled in her bad eating habit by providing the needed finances. (Apparently, the person had been helped repeatedly in the past without solving the basic problem.) So what should I do in obedience to God's will?

At our last class meeting as of this writing, this person announced that her immediate financial need had been met by help from group members (of whom I was secretly one) so that she can remain in her home for now. But how can this person's long-term needs be met? This problem is yet to be resolved. (Or may the situation be that God has herein provided the group with a way to show neighborly love on an ongoing basis so that there really is no problem? But would such enabling really be God's will?) Now let us see what circumstances unfold, trusting God to work his will in the matter as we continue to show love in ways that are pertinent.[62]

Let me further illustrate with another anecdote. Some years ago, I entered the Bolan Lake camp ground in southwest Oregon rather late in the day. I managed to find a campsite that had apparently just been vacated. It was near the lake shore and at the foot of a brushy rise that had a trail leading to a hidden campsite. As I began to set up my tent, suddenly there was loud and profane hollering emanating from the campsite at the top of the rise. It seemed that everyone in the campground gave their attention to the disruption. But I was the closest to the problem, certainly the nearest "neighbor." I felt led (by God), to promptly climb the trail to encounter the upset camper. In fact, he was very angry, claiming someone had stolen his coffee. He explained that he was homeless and had been seeking employment without success. In a reflex response, I told him that my master had instructed me to love my neighbor as myself and therefore I would give him half the money I had in my wallet, a rather modest sum. This caused a complete change in his demeanor. He accepted the money and thanked me. In returning to my campsite, I encountered other campers curious as to the outcome of my "foolish" reaction to the original hollering, expecting that I may well have been beaten up by the angry man. I explained to them the circumstance and my response and that all was now well. One of the camping families said that they were members of a nearby church. Soon they were taking food up the trail to the needy man. And others in the campground joined in, in helping and encouraging him, even inviting him to join them at their campfire. I had to leave early the next morning, so I do not know the final outcome. I had told the needy camper that I would be praying for him, including that he would soon get a job.

The Role of Prayer in Regard to God's Will

We note that Christ's request in "The Lord's Prayer" is that God the Father's "will be done." It seems that we, who are members of God's kingdom here on Earth, are prone to request God to do our will in many situations. How is this discrepancy to be understood and resolved?

Suppose, now, that the function of prayer is more to strengthen the Christian's faith than it is to control God. That is, God allows us, as members of his earthly kingdom, the privilege of participation in his actions. God intends an outcome and allows us to trust him for accomplishing his will even though it may not be what we had in mind. (See Matthew 6:8, John 15:7, I John 5:14.) If we choose not to participate in prayer, God yet accomplishes his intentions, but we miss out on the blessing of participation in his will. Should not, then, our petitioning prayers be primarily a seeking of God's will in our particular circumstances? And as he reveals his will to us, even through resulting events and circumstances, our prayers of gratitude and praise are likewise a recognition of and honor to his sovereignty. We are accurate reflections of his image. Let me here illustrate with an anecdote.

Though He Slay Me, Yet Will I Trust Him (Job 15:13)

In mid-July of 2010, I was ushering at the church my wife and I attend. On this occasion, I had the honor of seating a man and his wife who had been missionaries in France as well as previous leaders in our church. They had returned from their work in France due to the brain cancer that the wife now had. As they entered for seating, Lorelei leaned heavily on her husband's shoulder for support. She was now quite feeble and it turned out that this was her last time to attend a church service. Over several months prior to this event, there had been Christians from many places on the globe, including me, praying that God would heal Lorelei. But her health had continued to deteriorate. How then, was God to be honored in this circumstance?

I was able to seat the pair a few rows in front of my usher station and thus witnessed an amazing thing. Especially during the congregation's singing of the hymn, "Majesty, Worship His Majesty," Lorelei raised her feeble arms as high as she could in praise. And what an inspiration this was to those of us who were witness! In spite of the natural progress of the disease, Lorelei remained true to her faith and honored God's will in her life. She died just a few weeks later on August 20 of that year. And may her faithfulness be an inspiration to the readers of this account.

Is Answered Prayer Only Imagined?

How would a "prayer experiment" be relevant to this question? Would God not actualize his intended outcomes whether or not they are prayed for (experimental group) or not prayed for (control group)? God acts in nature by means of natural laws, the Bible is quite clear on this matter (e.g., Psalm 104). Science documents this process. But the Bible also holds that God may act in "improbable" ways as well. Unnatural healings would be examples. Obviously, these are improbable events only if they are rare within God's usual mode of action. And the divine purpose of such rare events must be our encouragement as God's control of nature is demonstrated in ways especially meaningful to us. Otherwise, he need only use natural process to actualize his intentions. Thus, miracles are phenomena quite distinct from the usual natural process, and thus, experiments designed to reveal them as natural process are misguided. Let me illustrate with an anecdote.

On June 24, 2008, a newcomer, Mark, walked into a Reasons to Believe group meeting in Oregon. This was his first time to participate in this group. He was recognized by a leader of the group, Gary, as a University of Oregon dormitory friend whom he had not seen since graduation some twenty years past. At that time, Gary had had some struggles in his faith. Mark, as a Christian brother, had been a real encouragement to Gary. Now, many years later, Mark apparently had questions about his own faith and suggested to the RTB group that he was considering conducting an experiment on the effectiveness of prayer as a test for evidence of God's existence. And here he met an old friend who was now willing to help him with his own struggles in turn. Was this coincidence alone or God's action demonstrated by a more pertinent "experimental design" than Mark had in mind?

Coincidence as a Mode of God's Action

I give two examples, both personal. The first illustrates answered prayer, the second an outcome not prayed for directly.

My first example is documented in letters that survive in my possession. They constitute correspondence between various participants in the events. I also have memories of my parents' verbal accounts given to me as I grew and matured in our family home. I begin with an account of the context that will provide deeper meaning.

My mother and father first met in Iraq as Christian missionaries under a mission which was independent of any church organization. They fell in love and were married in the early 1930s.[63] I was their firstborn (1934). In 1939, due to a digestive system malady he suffered, my father was advised by his doctor to take a period of rest from his stressful normal activates. He and my mother decided to "vacation" in nearby Lebanon. They left their rented home in Baghdad together with us, their three children, and travelled first to Damascus on June 1, 1939.

Meanwhile, the international situation was deteriorating due to the activities of Adolf Hitler. As a consequence, our family was denied reentry to Iraq. Even so, on September 22 of that year, my father attempted to reenter Iraq, primarily to recover belongings. After a six-day stay on the border at Rutbah, during which time he attempted to contact the American Vice Consul in Baghdad, he finally despaired of entry and returned to his family who had stayed behind. They now must work out alternate plans. It was soon decided that they would return to North America. Their finances were meager, but sufficient help was provided in a timely manner, in itself representing answered prayer.

They booked passage on the *SS Excambrion* and departed from Beirut on October 8th. Travelling west through the Mediterranean Sea, they made port at Marseilles, France. Here the ship was delayed about fourteen hours from the scheduled departure time. They then left to exit the Strait of Gibraltar and cross the Atlantic Ocean to America. Now, such a voyage was understood to be hazardous at that time. For instance, on September 24th of that year, three neutral merchant ships bound for the United Kingdom had been sunk by German U-boats (submarines).

After exiting the strait, and when they were at sea west of Portugal on October 25, 1939, they encountered wreckage debris of the British freighter, *Ledbury*. According to the calculations of the purser aboard our ship, the *Ledbury* had been sunk by the Germans some fourteen hours earlier on October 24th! Had it not been for our fourteen hour delay in Marseilles, the debris would almost certainly have included that of our ship, the *SS Excambrion*! My parents later learned that one of my mother's sister's prayer groups had been meeting in prayer for our safety, apparently exactly at the time of our delay in Marseilles! This often repeated story very much impressed me during my youth.

The *SS Excambrion* "cautiously" continued its voyage across the Atlantic, safely making port at Boston on November 2, 1939. One hundred and thirty passengers were safely delivered, among them our family and a Harvard professor, emeritus, Dr. Kinsopp Lake, a theological and archaeological scholar who had been supervising an excavation in eastern Turkey before his departure. Even so, we were convinced that it was my aunt's prayers for us, the humble and monetarily poor missionary family that was especially in concert with God's will.[64]

A Finely-Tuned Coincidence

On August 6, 1983, I was collecting mostly butterflies along a 9,000 foot high ridge in the Lemhi Mountains of Idaho. This was my first (and only) experience in this area which is near Meadow Creek Lake, southwest of Gilmore and to the south of the birthplace of Sacajawea of Lewis and Clark renown. My companion, Nelson Curtis, was in a nearby area collecting sulphur butterflies, which later served among the paratypes of *Colias meadii lemhiensis*, which subspecies he and a colleague described (and named) in 1985. But, I was alone on this ridge to the west where I had wandered without his knowledge.

I recall that the area is characterized by granite boulders, some very large, and by gnarled old pine trees thinly scattered on the lower slopes. The ridge I was on still had a snow pack sloping to the cliff-brink to my right. Figure 19 is a picture I took looking north from the location of this incident. The cliff below me dropped hundreds of feet to the floor of this remote valley which is some distance west of the road access area.

I walked along the ridge between the rather steeply sloping snow field to the northeast and the ridge crest to the southwest which was high enough to shade this snowpack below me. I recall that I had a specimen in my net so I got out my forceps to carefully handle it. But I accidentally dropped the forceps on the snowpack. To retrieve it, I stepped onto the snow, which to my surprise was coated with a crust of ice. Immediately, my feet slipped from under me and I accelerated toward the brink, feet first! Apparently, in spite of my frantic attempts, I could not even slow my speed, let alone stop! How, then, am I here to tell the tale?!

The Fine Tuning

This incident occurred at the right time of day so that the sun had topped the ridge, at least at the location where I was sliding, long enough to soften the snow pack crust for just a few feet above the brink. I was able to dig my heels into the zone of softened snow just in the nick of time to avoid plunging over the brink. I was then able to carefully crawl back to safety. I was certainly glad that God had willed to feed the local vultures by other means that day! And, again, the experience has been a significant reminder to me of God's care for even me.

I was at the exactly right location and time of day to avoid accidental disaster. Of course, God could have had the butterfly I caught fly somewhere else so that I would not have dropped my forceps. But then, I would not have this story to tell of his special care. And does it not serve to illustrate the fine detail of his involvement in nature and in the events of our lives? Keep in mind our thesis here, that natural order, allowing the laws that control the universe, is maintained by God's immanency. He controls the reality into which each probability wave collapses, thus determining the universal eigenstate of each time horizon. What other explanation is there? That I may, indeed, have gone over the cliff in another universe within an infinite multiverse? Theistic determinism is much more logical to this observer. If, indeed, somethingness is infinite, as would be required by an infinite multiverse, how can it not include infinite sentience, God? And such sentience, being infinite, must be immanent to all universes, including the one we are in. Thus, God is aware of even thought-prayers as well as in control of events, allowing free will and free running, but achieving his will if even by contingency. Again, see the treatment in earlier chapters of this work.[65]

But How Does Loving One's Non-Kin Neighbor Relate to the Survival of the Fittest (Favored) Principle?

The concept of "kingdom" implies social structuring. How then is God's kingdom of sentient, free-willed beings (we humans) "on Earth" unique? How can it thrive if it breaks the principle of survival of the more fit at the expense of the weak?

The Bible portrays God's earthly kingdom as structured by the relational motivation of love. Total love of the Lord, God, as king and the equal love of self and others with whom we have social connection (our neighbors) who may or may not be kin (Matthew 22: 37-39, Mark 12:30-31). In contrast to non-kingdom social systems, leaders are to be servants to the led. One's social siblings are to be valued on an equal basis to one's self rather than in a social hierarchy of importance. Jesus taught and modeled this social structuring. For example, when at a feast, he, the leader, washed the feet of the disciples, which traditionally was a servant's duty (John 13:4-5).

In natural, non-socionic social system structuring, the "fittest" dominate resource use and survive at the expense of the weak who are thus left without life-sustaining resource when the population is at the "K" level. If the system's members were to share equally, then the members of the whole population would weaken and likely die out. None would survive.

But note that our human social systems are essentially all socionic (as defined in chapter eleven.) That is, the members use resource to support a social system that involves more than the survival of its members. There is use of resource to support the socionics of the system itself, which in turn provides access to otherwise unavailable resource which, yet in turn, supports the population itself, as well as a variable "standard of living" of the system's members. And it is the use of this "excess," standard-of-living resource which is expendable without endangering the survival of the system and its members. (Of course, there would be a threshold beyond which more resource use would undermine the whole socionic system and cause its collapse along with the population it supports. And again, see chapter eleven as to how the "tragedy of the commons principle" is involved.)

How then, does Christ's admonition to share relate? It is evident that it is this latter resource use that Christ distinguished in his instruction to share in acts of love and thus in support of God's kingdom. The rich, young ruler who boasted of his careful fulfillment of the laws of Judaism and asked what might be missing in his qualification as a member of God's kingdom, was instructed by Jesus to sell his possessions and give of his wealth to the poor. He was told to then come and follow Christ and in so doing, to trust God for his survival needs (Luke 18:18-23). (However, evidently many of Jesus's followers had ways of making a living within the economic system of the day. So trusting God for one's needs certainly didn't condone laziness.) But not having the love that belonging to God's kingdom brings, the inquisitive, rich man walked away in disobedience. And it is possession of this love that Christ cited as evidence of kingdom membership. Love that is not otherwise found in human, non-kin social interaction. That thus depends upon "supernatural" endowment in those who obey Christ's (God's) command to repent and believe on Christ as savior and thus be "born again" into God's kingdom. Christ said that by their love his followers would be known (John 13:34-35). (And, of course, this love is expressed also in ways other than sharing of resource.) Herein, then, does God's kingdom come to Earth.

One may ask, then, can a person rich in worldly resource be in God's kingdom? We note that it is the attitude concerning the wealth more than its actual possession that is the problem. It is the "love of money" that is the sin. And this is what Jesus tested the rich, young ruler for. (How this relates to success in, for instance, a capitalistic economic system is a matter I will leave for others to elucidate.)

We observe in our world that human social structure generally is fostered by religion of many forms. That "Christianity" is no exception should not be unexpected. It may serve this function, as do other religions, even apart from participation in God's kingdom. Thus, those who pride themselves in strict observation of rules of conduct, etc., but lack the motivation of God-given love, may not serve as true witness to God's kingdom, much as shown by the rich, young ruler example.

Disagreement, Yet in Love, Not Enmity

There obviously continue to be those professing to belong to God's kingdom who hold to other perspectives of biblical interpretation than that proposed in this thesis. I am anxious to extend my love to them, even in disagreement. And I am always willing to "change my mind" on issues as deeper truth is revealed. I am encouraged by participation in forums in which issues such as dealt with in this thesis can be discussed in mutual love and attempted understanding. Again, it is not to be our perspective of, for instance, the mode of God's creation that is to be our identity as Christ's followers, but rather the mutual love that we hold and express. And as we express the love that God endows in us, that His Spirit enables, his kingdom comes to Earth as it is in heaven.

No Tragedy of the Commons in God's Kingdom

Note that the resulting social structuring, even though it is socionic, avoids the tragedy of the commons problem. A shared pasture would not be destroyed by overuse due to greed. Each user would love his neighbor as himself so that all would share equally in its produce and could recognize its limits. None would be rewarded for greed, rather all would have more modest but sustainable reward for cooperation. And if this principle were to hold in all areas that are commons, including even the reproductive rate of families, then the future of socionics would be bright. May God's kingdom thus come to Earth as it is in heaven.

And what of God's kingdom in heaven? How are we to participate? Where is heaven? What is it like? How are we to get there as kingdom members? What of life after death?

CHAPTER FOURTEEN:
Heaven: Evidences and Implications

Assuming the existence of heaven, how is it to be described? A perspective elaborated in, for instance, several recently published books holds that heaven is characterized by the lack of what we humans consider "bad" and an extravagant representation of what we consider "good." The Bible informs us that it is occupied by beings: God in trinity, as well as his angels, and by redeemed humans.

The Bible also indicates that there is more than one heavenly realm (e.g., 2 Corinthians 12:2). In this treatise, we have suggested that one of these realms, paradise, actually is what we term the "heaven brane," the brane universe counter to the "earth brane" (our universe) of the ekpyrotic model. We have not speculated on the other heaven or heavens. Perhaps they are other "slices" of the system, or perhaps they are more comprehensively inclusive, whatever that may mean.

Paradise

In the gospels of the Bible's New Testament, Christ informs the repentant thief hanging near him in crucifixion, that that day they would be together in "paradise." The inference is that, also upon their death, all humans who have been redeemed through repentance of the results of our fallen nature and acceptance of God's gift of membership in the heavenly kingdom by faith in Christ's provision of atonement, will likewise occupy paradise, at least until the eventual resurrection.

If, indeed, the heaven known as paradise is only a fourth spatial dimension from us in our universe, what would be the repentant thief's and his fellow, deceased and redeemed humans' present status? Let us use reasoned speculation on the matter.

First, what must be the contextual environment? If, indeed, paradise is the remnant Garden of Eden, then there must be plants present as well as "heavenly" animals such as Adam named. Adam and Eve ate of the plants in the Garden of Eden, implying that their bodies needed nourishment. This certainly implies that entropy is involved. Thus, this is in agreement with the ekpyrotic model.

If entropy occurred in Eden, then was there death? There certainly seems to have been at least the "death" of plant tissue. In Genesis 2:21, God makes garments of skin for disobedient Adam and Eve, implying that animals died in the process. (However, the skins could have originated in the earth brane to which Adam and Eve were promptly banished, according to our interpretation.)

If, indeed, the opposing universe brane of the ekpyrotic model is a heaven brane containing paradise, how is it characterized physically? Are there galaxies containing stars with planets? Is, then, the Garden of Eden on a planet counterpart to the planet Earth, but outside our universe brane? It is interesting that dark matter apparently is concentrated in the vicinity of the heavenly bodies of our universe. Does this suggest that there are corresponding heavenly bodies in the heaven brane if, indeed, dark matter is the gravity-causing matter of the opposing universe? And might the remnant Garden of Eden be on a planet corresponding to planet Earth? But what of a corresponding sun?

Referring to the status in The Holy City, Rev. 22:5 states, "And there shall no longer be night and they shall not have need of the light of a lamp nor the light of the sun, because the Lord God will give them light . . ." Note that gravitational attraction of a "heaven planet" to a sun (local star) is not excluded as a need. (Perhaps, though, the universe brane opposing ours in the ekpyrotic model has other laws of its physics so that galaxies are replaced by some other systems. If so, this certainly would allow us much speculation.)

Note that, according to the Bible, the heavens are not eternal. They had a beginning at creation (Genesis 1:1). At least a heaven (paradise?) as well as the earth (earth brane?) are to be eventually "recreated," this also in agreement with the ekpyrotic model. Apparently, angels are not eternal either, having also been created (in a context not well specified in the Bible). Nor is their bodily form (eye anatomy, etc.) nor the somethingness of which they are composed clarified in the Bible.

Our human bodies are, of course, made of the elements of the earth brane ("dust of the ground" in the concept of the ancient writers). But is there more to our existence? Of what would "spirit" be composed?

And what would be the relationship of human post-death existence, proposed in the Bible, to our pre-death existence? What would be the mode of information retention involved? (Note that there is a distinction in the Bible between pre-resurrection status and post-resurrection status. We are to be like the risen Christ in our post-resurrection status.)

Some have wondered why a resurrection would be needed at all if we are already in heaven! Indeed, if such a heaven is to be renewed (the heaven brane is to be recreated at a brane collision event), then we would have to be in a new bodily form, such as Christ's, to survive. Perhaps survival would be by entry into some other realm involving other dimensions, thus serving as a temporary abode. Earlier, we suggested that the New Jerusalem may serve this function as a "cosmic ark." Indeed, might the biblical account of the flood at Moses's time, quite functional in validating the historicity of the scripture in that it relates to flood accounts in other contemporary cultures, again be an "unwitting allegory" predicting this future complete eradication of life by the process involved in the brane collision splat event? This is to include an overwhelming "flood" of plasma ("water" to the ancients?). Here we only speculate, hopefully within reason.

Identity Preservation After Death and Decomposition

As individual humans, our personal identity or self-consciousness persists with time as long as our central nervous system, particularly our brain, remains functional. Our self-awareness is temporarily lost during sleep and accidental or medically induced unconsciousness. But it resumes upon regaining consciousness as long as our brain is intact in its order and function. But what of our self-identity upon disruption of this physical order and function such as at our death? Is there evidence of a "spirit" that persists such deterioration? And, if so, what would be the nature of such spirit?

Years ago, when I was a graduate student at the University of Washington, a classmate survived a severe traffic accident in which he suffered extensive brain damage. He eventually recovered physically, but, at least to me, he seemed to have a different personality, to be an essentially different person. To me this evidenced the dependence of us, as individual personalities, upon normal brain function. On the other hand, I remember well the look of awareness in my brother's eyes when I last saw him, before he died of the brain cancer which had completely immobilized him otherwise.

But even our normal brain function and personality change with time. I wonder if I were as I am now to somehow meet "myself" from years past, if "we" would not function as separate individuals. And what of the result of certain brain surgery in which the connection between the brain's hemispheres is severed, creating two virtually independent "selves" within the same head?

How then can one's personal identity and awareness persist after death and brain decomposition? Of what might one's supposed "spirit" be composed and how could it relate to brain function in life and persist after death?

Answers From "Ekpyrosis": Do We Have Heavenly Avatars (Spirit-bodies)?

2 Corinthians 5:4-5: "For while we are in this tent, we groan . . . we [rather] . . . wish to be . . . clothed with our heavenly dwelling so that what is mortal may be swallowed up by life. Now it is God who has made us for this very purpose and has given us the spirit as a deposit guaranteeing what is to come."

Quantum entanglement amongst probability waves apparently accounts for coordinated decoherence into the universal eigenstates that characterize each time horizon in our universe.[66] Could it be that such probability waves actually extend through the separating dimension and into the heaven brane? And if entanglement is likewise not limited to our dimensions, could not information be shared between the universes, perhaps by coordinated collapse points at each decoherence? If so, would not eigenstates be likewise extended? And might this not allow counterpart beings in both universes? Do we have heavenly avatar counterparts? (If so, would my avatar really be "me"? Note that it must exist even while I am living in my earthly "tent.")

Perhaps this seems to be unreasonable speculation, but is it not more parsimonious than the postulated infinite replications of me of the multiverse hypothesis?! There it would be the exhaustion of chance that mandates my multiple existence. In our present hypothesis, there is shared information that allows at least one "me" duplicate.

If this speculation is actual, then your avatar is very close to you, indeed, being only the tiny distance away of the fourth spatial dimension separation of the branes. And could it not involve a duplication of every molecule and physical structure of your composition? (See "mirror particles" on Google.) Perhaps even your thoughts are shared, although, at least usually, your awareness is somehow limited to the earth brane realm.[67]

Perhaps there is some credibility to the "out of body" experiences that have been reported. If the avatar were to leave the vicinity of one's dying body and thus not participate in the dying, then perhaps its awareness could be from a nearby location. And, not being limited by gravity in our dimensions, could it not be held by the gravity of the other brane, which might allow its perspective to be that of a location "suspended in mid-air" as has apparently been reported? (Remember that we earlier suggested such a phenomenon for the reported ascension of Christ. A difference being, however, that our avatar would be existent only in the opposing brane and hence not visible to observers from the earth brane perspective.) There is still the question of avatar awareness of earthly context if only gravitons, and not even photons, can pass through the fourth spatial dimension. How could such a suspended avatar see what was going on? Perhaps entanglement is somehow involved here also?

Plant and Animal Avatars?

If we humans have heavenly avatars, what would preclude such avatars for animals and even plants? Would there be physical features in the heaven brane that correspond to those in this brane as well?[68] From the perspective of a heaven brane Garden of Eden, perhaps there was an earthly garden avatar. I will leave speculation on such possibilities and their implications to others. One keeps in mind that immanent God is involved in all aspects and certainly can modify events and context conditions.

(Perhaps by now you, the reader, realize that much of my motivation in this thesis is to plant ideas, hoping that they will germinate and grow in your fertile mind. Hopefully, profound concepts will result and be shared. And I trust that, in the process, biblical relevance is honored.)

Alternate Information-Preservation Models

One might speculate that, if the avatar model were not actual, there could be other means of information preservation allowing afterlife. For instance, if somehow universal eigenstates, or at least the information they contain, are preserved, could God not access them in the future and reinstate at least aspects of them as reality? Apparently, wave function collapse, and thus each eigenstate is irreversible, and thus cannot revert to previous states.[69] However, subsequent eigenstates apparently usually resume the information status of those previous even though the intermittent wave function status allows the probability of deviation to what could be disruptive, even disorderly reality. Does this not imply that the information of each eigenstate is somehow persistent? How else could the next usually be consistent with the previous? And would not God, in his omniscience, be aware of such information and, in his omnipotence, be in control of the process (and thus able to sustain the order of the universe)? And could not God then use the information that gives identity to our individual being to somehow "recreate" us in heaven upon our earthly death? However, it seems that 2 Corinthians better fits the avatar perspective, which may thereby be the expression of God's information retention, at least of us personally.[70]

In this context, the biblical message of God's forgiveness of our sinful condition upon our repentance and obedience, has special significance. Does God alter the eigenstate records to obliterate our sinful past and replace them with Christ's redemptive provision? But, it seems, this would constitute falsification, and God is "truth." So it may rather be that God substitutes the truth of Christ's redemptive provision for our actual record and on this basis we qualify as reinstated representatives of Adam and Eve in their pre-fall innocence and willful obedience. We thus become accurate images of God and enter his kingdom, including its heavenly aspects.

Herein is another possibility. May not information, including that of our detailed composition and mental awareness, somehow be preserved in holographic systems such as have been proposed for even the retention of information from entities that enter black holes?[71] Again, substitution of Christ's provision for our redemption would be required for entry and persistence in God's kingdom.

Why Shouldn't the Universe Exist? A Concept Review

It would seem to be such a waste of existence if we are not to somehow persist our earthly death. Would there not, indeed, be no reason for the existence of the universe? Of absolute somethingness? Of God? Then why not absolute nothingness? But, in fact, there *is* somethingness and we *do* exist as witness. Why, if not by design, by theistic determinism?

Existence in the Heaven Brane: Are There Shadows in Heaven?

Earlier, we referred to the biblical statement that in heaven there is no need for the light of the sun. Rather "light" is provided by God himself. One may ask, is this light composed of photons and also with wave function or may it be unique from the light that our science has described? Could it be that somehow in the heaven brane the light that illuminates all beings comes from all directions so that there is thus no night nor even a shadow? If God is omnipresent in the heaven brane and emits light constantly and without depletion, then it must consistently come from all directions. And suppose that such light can impart replacement of any energy lost due to entropy, here ongoing as compared to energy replacement at subsequent brane collisions. Suppose also that such light is of a nature that allows all living things to intake it directly and thus such life is immortal and there is no degradation in form or function. [Could the "negative gravitational energy" postulated by Steinhardt and Turok, in *Endless Universe* (New York: Doubleday, 165, 191-192), for recharging the ekpyrotic model at each brane collision be of similar source? Indeed, it is considered of infinite supply. Holding that God creates by such collision implies that it is.]

If such Godly light directly energizes heavenly living things, then perhaps "eating" is to supply heavenly nutrients, consisting of the equivalent of earthly elements and molecules. (Adam and Eve ate fruits. But fruits are reproductive in function, at least here on Earth. So do plants reproduce in heaven even though, according to the Bible, humans do not?) Presumably, however, some form of "waste" would result from entropy in the living process.

Would breathing oxygen be necessary? Indeed, is air as we know it even present? (One could argue that sound could be transmitted other than by airborne sound waves.) Perhaps sentient beings could transmit and receive messages without the need of sense receptors such as earthly animals require. One might expect that at least angels are composed of subunits quite different from cells as we know them. Would they even need circulatory systems with the entailed blood? Such is the adventure of reasoned speculation!

We note that active existence depends upon an increase in entropy. Things could just "be" but not "do" in the absence of increasing entropy. The Bible is quite clear that heavenly beings "do" as well as exist. Their activities therefore increase entropy in the heaven brane and, of course, entropy accumulates in our brane as well. And the ekpyrotic model explains the accommodation of such entropy in both branes. It is kept diluted by the continued expansion of the branes during the intervals between the brane collisions. This expansion has been documented in our brane universe lending credibility to this theory. And it seems that it would be of necessity in heaven, suggesting that the idea that the brane opposing ours is, indeed, at least a heaven, is certainly credible. (Presumably God-light compensation for entropy would be appropriately limited.)

Any evidence of the validity of the ekpyrotic model such as may be detected by space probes studying cosmic gravitational waves (Steinhardt and Turok, ibid., 197-198) is certainly pertinent to the perspective suggested in this thesis. One trusts that such evidence is soon found and reported. If instead, the evidence further supports the inflationary, big bang theory and disqualifies the ekpyrotic model, then the issues dealt with in this thesis would have to be reconsidered. Of course, there could still be a heaven, but its location and aspects of its nature would remain unknown (in our earthly context) at least until further advances in cosmology.

Implied Consequences of the Fate of Earth and Heaven

Modern cosmology recognizes that, in time, this universe will expand and dilute to a state of high entropy in which galaxies, stars, and planets, including planet Earth, will no longer persist. The Bible declares that the present heaven and "Earth" (surely more inclusive than the planet) will "pass away."(But to be replaced. See Revelation 21:1.) And yet the Bible promises that we will exist into eternity, that is, forever beyond even our presence in paradise, the first heaven (e.g., John 11:26). How can this be?

The Resurrection

I Corinthians 15:40 There are heavenly bodies and there are earthly bodies.

I Corinthians 15:44 b If there is a natural body, there is also a spiritual body.

I Corinthians 15:53 The perishable must clothe itself with the imperishable . . .

I Corinthians 15:52 In a flash, in the twinkling of an eye, . . .the dead will be raised imperishable, and we will be changed.

The Bible declares that Christ was raised from the dead and that his resurrected body is a precursor to that of each human to be likewise resurrected at his return to this Earth from heaven where he now is in his resurrected body. All humans are to be resurrected, both the "redeemed and the wicked" (Acts 24:15). The implications are that the spiritual bodies of redeemed humans are in paradise in the heaven brane whereas the spiritual bodies of the wicked are elsewhere, but apparently also in what we have termed the heaven brane. But both the earth brane and the heaven brane are to be destroyed and then made new, both a biblical perspective and the perspective of the ekpyrotic model. Thus, the resurrection of all humans into a form that is not restricted to either brane is logically required. How might this be? What might the resurrection process entail?

In our earlier discussion of the possible process involved in Christ's resurrection, we hypothesized that, as Christ's earthly body was three-dimensional, if it were somehow united with the three dimensions of the opposing universe, the heaven brane, that the combination would then be six-dimensional and thus able to pass through the fourth dimension that separates the counterpart universes of the ekpyrotic model. (It seems that such a process would necessitate a heavenly avatar for Jesus as for us.)

As recorded in the New Testament, the resurrection of Jesus Christ involved only his body, not his garments. And apparently, the surviving molecules making up his body were involved in that they did not remain in the tomb. And his resurrected body is described as retaining the damage, at least of the nail prints, from his crucifixion. (He instructed doubting Thomas to inspect his hands for this damage.)

Mummies Versus Ashes

In order to conform with Christ's resurrection process, should we be mummifying our dead as did the Egyptians? What of cremation and the spreading of the resulting ashes? How might the resurrection process be independent of such circumstances?

Now if our heavenly avatars preserve the organizational information upon which our identity depends, could they not serve as "templates" for the subsequent reorganization of the earth brane components of the six-dimensional resurrected beings we are to become? Perhaps such a combination of components would involve translocation of the needed open-ended quantum strings as their probability waves collapse appropriately under God's control. Their source, either from remaining mummies or from scattered ashes or elsewhere would thus not be limiting.[72] One might even generalize the source as "the dust of the Earth."

Then, as the earth brane strings match end-to-end with the heaven brane strings of the heavenly avatars, perhaps they then form string rings that, like gravitons, could pass through the fourth dimension as well as likely have other "supernatural" properties. In this way the status of our personal identities would be reestablished in the earth brane even if long after our physical death. And, as suggested earlier, our being would persist the destruction of heaven and Earth (the next ekpyrotic collision phase) into the new heaven and new Earth phase. Thereby we then will have access into both universe branes by means of our six-dimensional bodies.

The Bible states that the new Earth (brane?) will no longer be "under the curse." It would seem that this means that there would be different physical laws, perhaps now like those of the heaven brane. Hence, the increase of thermodynamic entropy upon which action depends (Carroll, *From Eternity to Here*, New York: Dutton (Penguin Group) 2010; p. 391) may be compensated by "God light" as we postulated for the former heaven brane.

And What Will We Do for Eternity?

Doing results in change. Does eternal "doing" result in infinite change? But if perfection is the starting condition, why change at all? But it seems that perfection must necessitate more than just "being." Somehow doing without disrupting perfection must be possible in heaven. And how is perfection to be defined?

God's will is perfect by definition. Thus, doing God's will in obedience and with the devotion of loving God with "all our heart, all our soul, and all our mind" (Matthew 22:37) will constitute the "doing" of the redeemed. And thus such doing will not change perfection. And only those so motivated can participate or heaven (old and new) and the new Earth would be imperfect.

God has provided us, finite sentient beings, with free will and a context that requires faith in its exercise. One concludes that our existence in such a context has been God's intention from eternity. The exercise of our free will to obey God's will thus fulfills this intention. And as we do this in the earth brane context, our free will can be exercised "freely." Thus, we can choose to either obey or disobey in faith in a context where the consequences are not established other than by faith. And it is in this context that God's perfection is honored, that we become truly in God's image and thus reflect his sentience, his will. This opportunity is lost upon our entry into the afterlife where faith will no longer be required in choice-making.

Conclusion

What then will the redeemed do in heaven for eternity? The answer: God's will. And those who choose to reject doing God's will in this earthly context where faith is required will not be compelled to do it in the afterlife, but will be expelled to "hell" according to the Bible. Here their choice to reject God's will, will be forever "honored."[73] And the perfection of heaven will not be violated.

For the redeemed, doing God's will for eternity should be with utmost fulfillment. What might be God's intentions for subsequent cycles of the ekpyrotic system? What roles might the members of God's kingdom have in these events? Is there, indeed, a multiverse? What sort of dimensionalities would be involved? If, indeed, the resurrection bodies of the redeemed are six-dimensional, then would they not be able to pass into various universes? It certainly would seem that the afterlife will not be boring for the members of God's kingdom.

Decisions, an Exercise of Free Will

And so, dear reader, if you have not already done so, it is with utmost concern and love that I urge you to heed God's invitation communicated through the Bible, his Word, to enter the Kingdom of Heaven. Recognize and repent of the fallen condition of sinful disobedience we all are born into. Obey God's command to believe in the atoning grace provided by Christ's sacrifice for your salvation (John 3:16). As you seek, you will find. God's immanent presence, through the Holy Spirit, will work in your "heart" to confirm your spiritual rebirth. The Bible proclaims that even the angels in heaven will rejoice as you thus enter God's Kingdom. And I will get to meet you and enjoy your company forever!

And if you so choose to reject this invitation, the Bible proclaims that God will exercise the justice on you that we all deserve in our unredeemed status. For me, that will bring much sorrow that you missed the opportunity. My intent in composing this book will be unfulfilled in your regard.

Appendix I. Macroevolution Explained

There apparently is some confusion in the meaning of the term "macroevolution." The following two perspectives are here considered:

1. Change in a population beyond that of "gene shuffling" via sexual reproduction in response to change in adaptive pressure (microevolution) is apparently considered to be macroevolution by some. Such change usually requires mutational alteration necessary for the production of alleles new to the gene pool of the population. Or, in some cases, it may be accomplished by cross breeding with other related species so as to access their gene pools. In that it requires more time and the existence of novel genes, it may thus be distinguished as "macroevolution."
2. Change in fundamental organizational patterns that delineate higher categories, such as phyla is another macroevolutionary concept. Such features as body symmetry and digestive system design are examples. Such basic patterns are apparently determined at ancestral stages where the features are relatively simple so that "microevolutionary" (here more inclusive in meaning) process has macroevolutionary potential. The following considerations attempt to illustrate the principle:

How Do New Phyla Originate?

Embryonic progression from a single cell to multi-celled tissues and organs involves patterns that characterize different groups of organisms. Although these patterns may be modified even during embryonic stages, such as the effects caused by increased yolk storage, generally they are considered to reflect chronological modification as complexity increases in ancestral lineages. (Hence, the reflection "ontogeny recapitulates phylogeny" long recognized by biologists.) In that such basic patterns are determined early in phylogeny, it is not logical that new species at subsequent time horizons would shift from one basic pattern to another. Let us here illustrate the principle with an example: the divergence of the Protostomes and the Deuterostomes.

As the names imply, the mouth in one group derives from a primitive opening whereas it is a secondary opening in the other. Development progresses from the fertilized egg cell to a hollow ball of cells, the blastula. Then there is an infolding of the surface layer of cells to form a double-layered stage with an opening, the blastopore, into the new internal cavity thus formed. This new cavity functions in primitive organisms as a digestive chamber. This allows use of multi-celled food units so that food vacuoles of individual cells are no longer limiting to food unit size. However, the new opening must serve both for intake of the food units and for the expulsion of undigested waste at this organizational level. This two-way movement pattern precludes regional specialization in the digestive system. This limitation is mediated in two ways, each characterizing the major phylogenetic branches cited above.

In the Protostomes (including the Mollusks, Annelids, and Arthropods) a new opening forms that functions as an anus, thus allowing a one-way flow of contents and the potential for regional specialization of the tract. But in the Deuterostomes (e.g., Echinoderms and Chordates), it is the new opening that serves as the mouth, and the original opening that functions as the anus. This also allows one-way flow but in the opposite direction. Obviously this is a fundamental organizational difference. We, as Deuterostomes, are organizationally backwards to, for instance, earthworms and beetles.

One thus does not expect an insect species to give rise to a mammal! Macroevolution at this level began with "microevolutionary" stages at the appropriate ancestral levels.

Could new phyla still originate? What would be the potential if, for instance, the Porifera (sponges) which are organized as filter feeders, still restricted to food vacuole digestion, were to modify so as to form an internal digestive cavity, thus allowing multi-celled units to serve as food? Apparently, there, indeed, is a deep sea sponge group, mostly in the family Cladorhizidae, that is carnivorous on multi-celled prey. Apparently, mobile cells congregate around a prey in such a way as to enclose it, providing a chamber in which it can be digested by secreted enzymes. If such a new process eventually results in a novel organizational pattern in which a permanent digestive system arises, thus providing the potential for other adaptive modifications, then would not a new phylum have evolved?

For further detail and illustrations see appropriate textbooks (or Google). Ruppert and Barnes in their *Invertebrate Zoology* text (Fort Worth: Saunders, 1994) give good coverage of such issues. Figure 5-19, on page 198, of the 6th edition well illustrates the developmental patterns described above.

Appendix II. Theistic Determinism, A Summary Statement

If Heisenberg uncertainty is actual, then how can laws determine consistency? Rather, consistency must determine laws and this resulting in emergent order. How, then, can this be other than by divine causation? Theistic determinism. Thus, the laws of nature are God's laws and the dynamics of nature are God at work. It is thus natural for God to work through nature. God is both natural and supernatural, immanent and transcendent.

Appendix III. Does Absolute Nothingness Exist?

On page 170 of *A Universe from Nothing*, Lawrence Krauss defines "nothing" as even less than "empty space," so that it would refer to the lack of dimensions (especially those of space as we know it). This is essentially my concept of "absolute nothingness" that I hold to be logically non-existent. I hypothesize, rather, that the lack of dimensions would necessitate the existence of "everything" coincidently (everythingness rather than nothingness). This may be in a primary state of potentiality, but would require a "somethingness" context which must be infinite. I, of course, define such somethingness as God. (That absolute nothingness could contain such potential serves as a null hypothesis falsified by the philosophical conclusion that such nothingness cannot exist.) Potentials in "God's mind" would then be actualized in a context of the "creation" of the appropriate dimensions.

Bibliography

Barbour, Ian. *When Science Meets Religion*. New York: Harper Collins, 2000.

Brennan, Richard. *Heisenberg Probably Slept Here: The Lives, Times, and Ideas of the Great Physicists of the 20th Century*. New York: Wiley, 1997.

Badyaeu, Alexander V. "The Beak of the Finch: Coevolution of Genetic Covarience Structure and Developmental Modularity During Adaptive Evolution," *Philisophical Transactions of the Royal Society*, (April 2010): vol. 365, no. 1543, 111-1126.

Carroll, Sean. *From Eternity to Here: The Quest for the Ultimate Theory of Time*. New York: Dutton, 2010.

Clegg, Brian. *The God Effect: Quantum Entanglement, Science's Strangest Phenomenon*. New York: St. Martin's Press, 2006.

Davies, Paul. *Cosmic Jackpot*. Boston: Houghton Mifflin Harcourt, 2007.

Faris, Robert. *Social Psychology*. New York: The Ronald Press Co., 1952.

Ferris, Timothy. *The Whole Shebang: A State-of-the-Universe (s) Report*. New York: Touchstone, 1997.

Gleick, James. *Chaos: Making a New Science*. New York: Penguin Books, 1987.

Goodall, Jane. *In the Shadow of Man*. Boston: Houghton Mifflin Co., 1988.

Gould, Stephen J. *Wonderful Life: The Burgess Shale and the Nature of History*. New York, London: W.W. Norton, 1989.

Greene, Brian. *The Elegant Universe*. New York: Vintage Books, 2000.

————.*The Fabric of the Cosmos*. New York, Toronto: Alfred A. Knopf, 2004.

————.*The Hidden Reality*. New York: Alfred A. Knopf, 2011.

Gribbin, John. *In Search of Schrodinger's Cat: Quantum Physics and Reality*. New York: Bantam Books, 1984.

Guppy, Crispin and John Shepard, *Butterflies of British Columbia*, Vancouver, Toronto: UBC Press, 2001.

Hawking, Stephen. *A Brief History of Time*. New York: Bantam Books, 1990, 10th anniversary edition, 1998.

————.*The Grand Design*. New York: Bantam Books, 2010.

Hardin, Garrett. "The Tragedy of the Commons," *Science* 162, (1968): 1243-1248.

Honda, Keiichi. "Larval Osmeterial Secretions of the Swallowtails (Papilio)," *Journal of Chemical Ecology* 7, no. 6 (1981).

Hummel, Carles E. *The Galileo Connection*. Downers Grove: Intervarsity Press, 1986.

Kaku, Michio *Hyperspace*. New York: Anchor Book: Doubleday, Oxford University Press, 1994.

————*Parallel Worlds*. New York: Anchor Books (Random House), 2006.

Krauss, Lawrence M. *A Universe from Nothing, Why there is Something Rather than Nothing*. New York: Free Press, 2012.

Lamoureux, Denis O. *I Love Jesus and I Accept Evolution*. WIPF STOCK, Eugene, Oregon, 2009.

Madar, S. S. *Biology*, 10th ed. New York: McGraw Hill, 2010.

Marean, Curtis W. "When the Sea Saved Humanity," *Scientific American* (August 2010): 55-61.

Murphy, Nancy, Robert Russell, William Stoeger, eds., *Physics and Cosmology: Scientific Perspectives on the Problem of Evil*. vol 1, Notre Dame: Vatican Observatory Foundation, 2007.

Newberg, A., and M. R. Waldman. *How God Changes Your Brain?* First ed. New York: Ballantine Books, 2009.

Nichols, Terence L. *The Sacred Cosmos: Christian Faith and the Challenge of Naturalism.* Grand Rapids: Brazos Press, 2003.

Parker, Andrew. *In the Blink of an Eye.* Cambridge: Perseus, 2003. (Deals with the cambrian explosion.)

————*The Genesis Enigma: Why the First Book of the Bible is Scientifically Accurate.* New York: Plume, Penguin Group, 2010.

Peacocke, Arthur *Evolution the Disguised Friend of Faith?* Philadelphia, London: Templeton Foundation Press, 2004.

————, with Philip Clayton *All that Is: A Naturalistic Faith for the Twenty-first Century.* Minneapolis: Fortress Press, 2007.

Peters, Ted, Robert Russell, William Stoeger, eds. *Resurrection: Theological and Scientific Assessments.* Grand Rapids: W. B. Erdmens, 2002.

Polkinghorne, John. *Belief in God in an Age of Science.* New Haven, London: Yale University Press, 1998.

————. *Quarks, Chaos and Christianity: Questions to Science and Religion.* New York: The Crossroad Publishing Company, 1994, 2005.

————. *Science and Providence: God's Interaction with the World.* Philadelphia, London: Templeton Press, 2005.

————. *Quantum Physics and Theology: An Unexpected Kinship.* New Haven: Yale University Press, 2007.

Randall, Lisa. *Knocking on Heaven's Door.* New York: Harper Collins Publishers, 2011.

Ratzsch, Del. *The Battle of Beginnings.* Downers Grove: Intervarsity Press, 1996.

Reichley, A. James, et al. *Religion in American Public Life.* Washington D.C.: Brookings Institute Report, 1985.

Richardson, Don. *Peace Child.* Ventura: Regal Books, 1974.

Ross, Hugh. *The Fingerprint of God*, 2nd ed. Orange: Promise Publishing Co., 1991.

————. *The Creator and the Cosmos.* 3rd ed. Colorado Springs: Navipress, 2001.

————. *Why the Universe is the Way It Is.* Grand Rapids: Baker Books, 2008.

Russell, Robert J. *Cosmology, from Alpha to Omega: The Creative Mutual Interaction of Theology and Science.* Minneapolis: Fortress Press, 2008.

Sire, James W. *The Universe Next Door.* 2nd ed. Downer's Grove: Intervarsity Press, 1988.

Stebbins, Ledyard, *Processes Of Organic Evolution.* Englewood Cliffs: Prentice Hall, 1966.

Steinhardt, Paul, and Neil Turok. *Endless Universe: Beyond the Big Bang.* New York: Doubleday, 2007.

Walton, John H. *The Lost World of Genesis One: Ancient Cosmology and the Origins Debate.* Downers Grove: Intervarsity Press, 2009.

Weiner Jonathon, *The Beak Of The Finch.* New York: Vintage Books, 1994.

The Bible. All Bible quotes used here are from: *The Holy Bible, New International Version*, Copyright 1973, 1978, 1984 by International Bible Society, used by permission of Zondervan Publishing House, Grand Rapids. All rights reserved.

Illustrations

Figure 1: Babylonian Brick

Figure 2: Middle Age View of Heaven and Earth

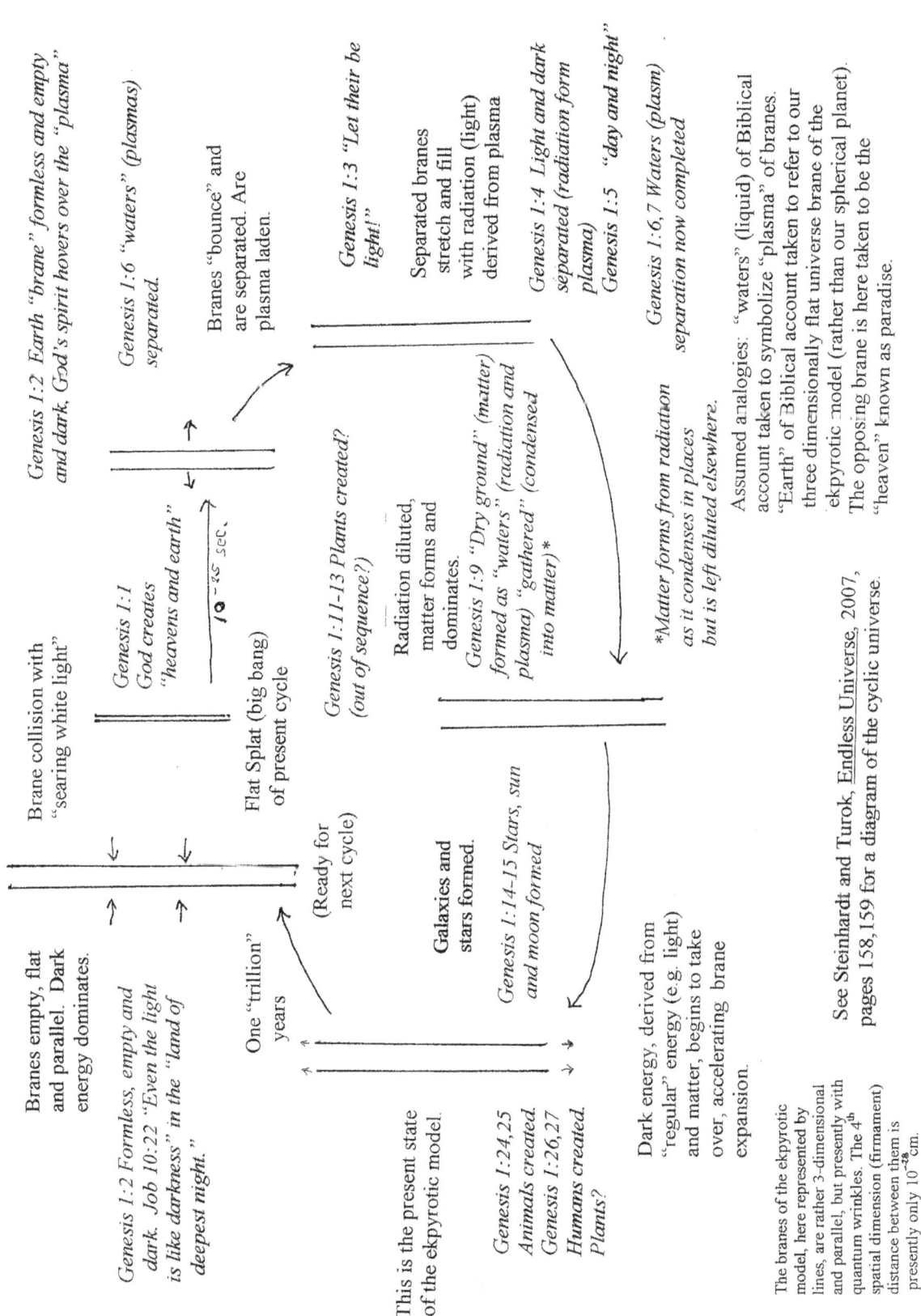

Figure 3: Genesis One Compared to the Ekpyrotic Model

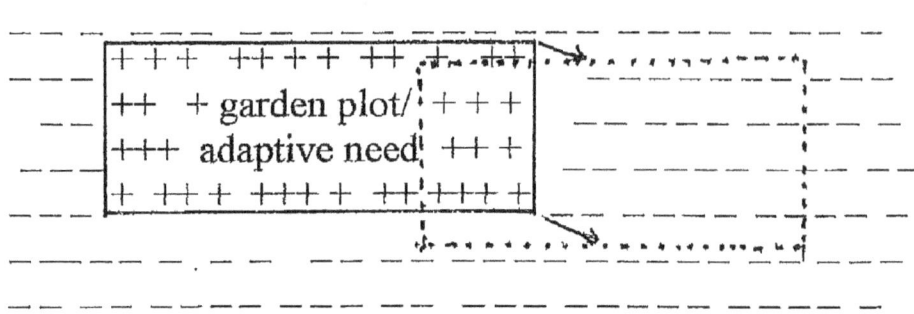

+, ⇀: raindrop target/genetic mutation

+: "useful" from relative perspective
⇀: "wasteful" from relative perspective
versus
No waste from God's infinite perspective

Figure 4: Relative Versus Infinite Perspectives (Garden plot shift)

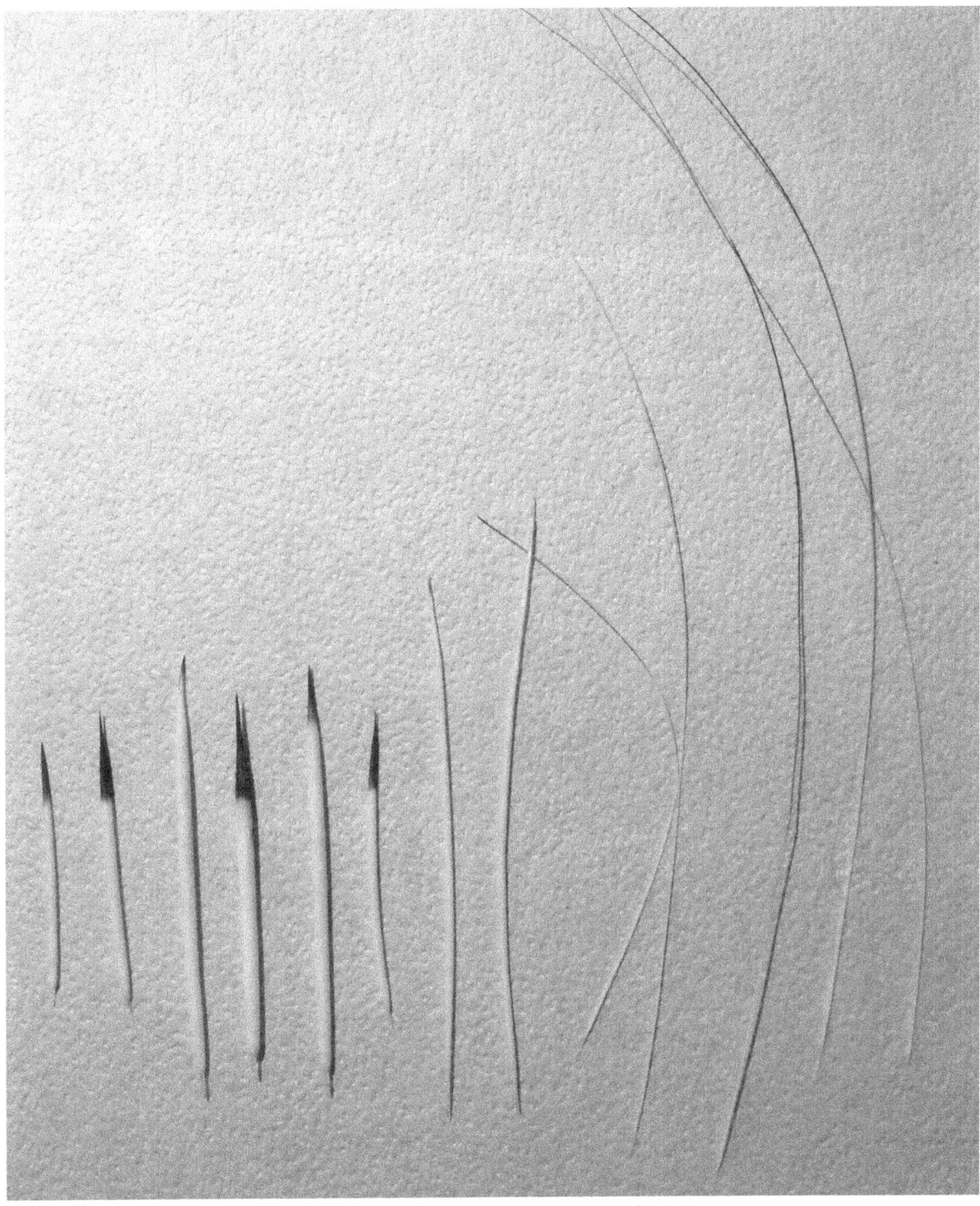

Figure 5: Porcupine Quills Versus Hairs

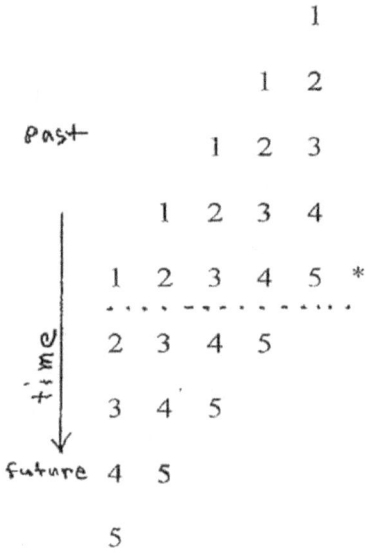

past

time

future

```
                        1
                    1   2
            1   2   3
        1   2   3   4
    1   2   3   4   5   *
    . . . - - . . . . . .
    2   3   4   5
    3   4   5
    4   5
    5
```

* Present time horizon

Stage 1 (see fig. 7)

Stages 3,4 (see fig. 11)

Stage 2 (see fig. 10)

Stage 5 (see fig. 9)

Figure 6:   Stages of Speciation: Timelines Versus Present Time Horizon

Figure 7: Stage One: *Papilio indra*: Subspecies Examples

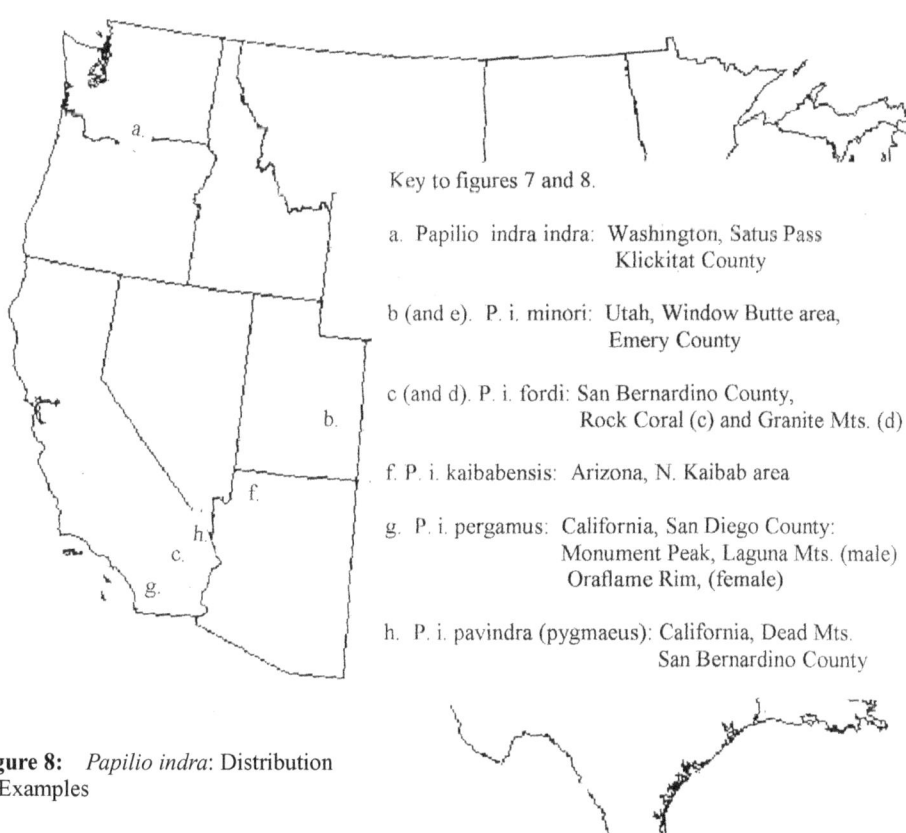

Key to figures 7 and 8.

a. Papilio indra indra: Washington, Satus Pass
Klickitat County

b (and e). P. i. minori: Utah, Window Butte area,
Emery County

c (and d). P. i. fordi: San Bernardino County,
Rock Coral (c) and Granite Mts. (d)

f. P. i. kaibabensis: Arizona, N. Kaibab area

g. P. i. pergamus: California, San Diego County:
Monument Peak, Laguna Mts. (male)
Oraflame Rim, (female)

h. P. i. pavindra (pygmaeus): California, Dead Mts.
San Bernardino County

Figure 8: *Papilio indra*: Distribution
of Examples

Figure 9. a. Papilio eurymedon Polk County, Oregon
b. Papilio rutulus male: Lane Co. Oregon, female: Clakamas Co. Oregon
(males on right, females on left)

Figure 10. a. Parnassius smintheus sternitzkyi   Siskiyou Mountains, SW Oregon.
b. Parnassius smintheus olympianna   Olympic Mountains, Washington
c. Parnassius clodius   Polk Co., Oregon
(males on right, females on left)

Figure 11. Papilio machaon complex examples.

a. P. machaon, Israel, male. b. P. machaon, France, male and female.
c. P. m. aliaska, near Fairbanks, Alaska, male.
d. P. m. pikei, Clayhurst, British Columbia, male and female
e. P. machaon times P. zelicaon, natural hybrids, Pink Mt., B.C., male and female
f. P. machaon times P. zelicaon, natural hybrids, Nordegg, Alberta, male and female
g. P. machaon times P. zelicaon, lab. hybrid, B.C. times Oregon zelicaon

Note. I am missing P. m. hudsonianus. For illustrations of this subspecies, Google "Papilio machaon hudsonianus" and choose option "pinned specimens, Butterflies of America". (The black forms indicate likely gene flow from populations to the east.)

Figure 12. Alexanor, a stem form.
a. Papilio machaon, Israel
b. Papilio machaon, France
c. Papilio alexanor, Israel
d. Papilio rutulus, Washington state
e. Papilio multicaudatus (small) Oregon
Although alexanor, rutulus and multicaudatus

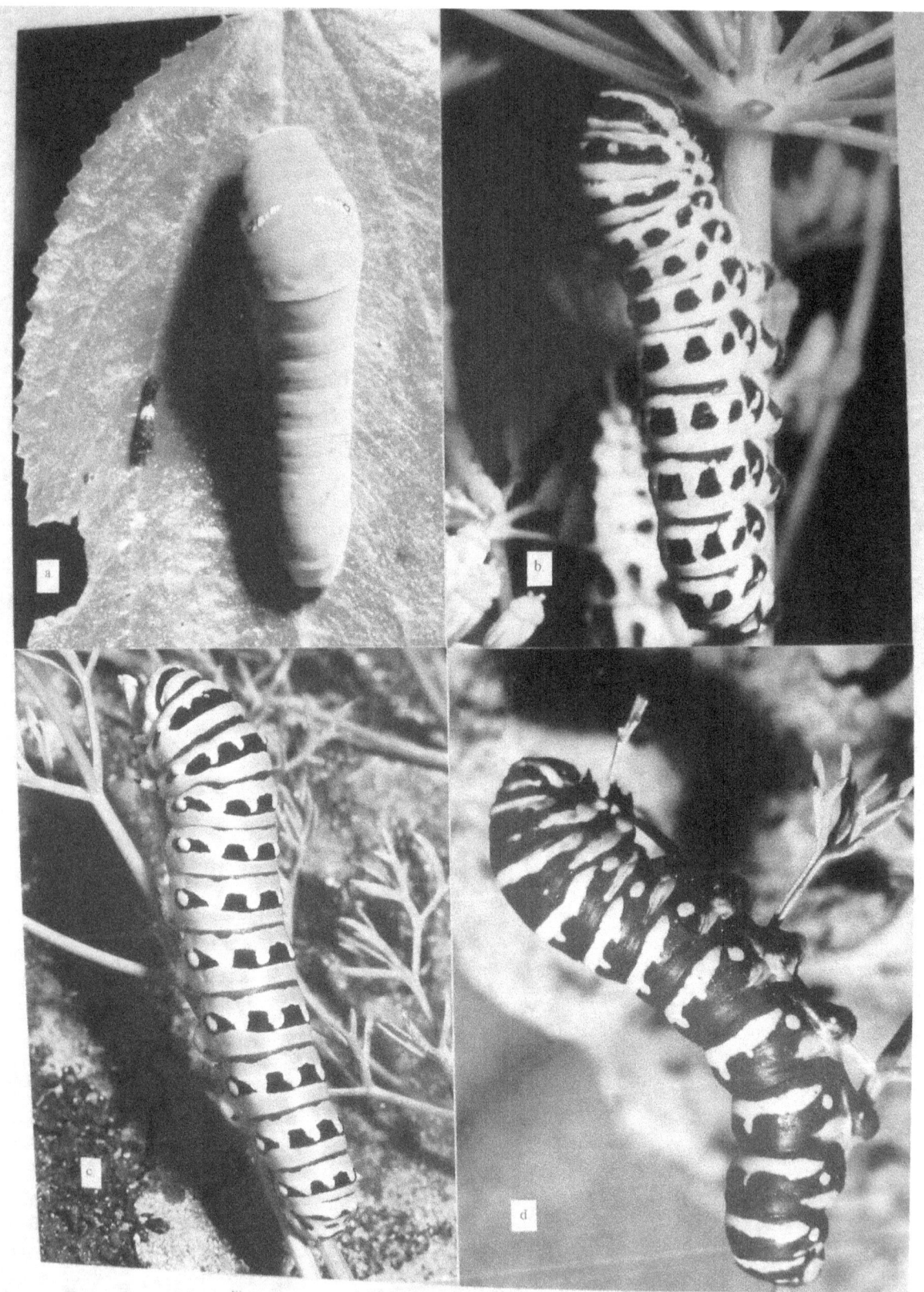

Figure 13. Larvae Compared
a. Papilio rutulus, early and late instars b. Papilio alexanor, mature larva
c. Papilio zelicaon, mature larva. d. Papilio indra, mature larva

94

Figure 14.
a. Papilio zelicaon. Male:upper, female lower (Oregon)
b. Papilio bairdii oregonius. Female:upper, male lower (east Oregon)
c. Papilio bairdii dodi. Female:upper, male lower (Cut Bank, Montana)

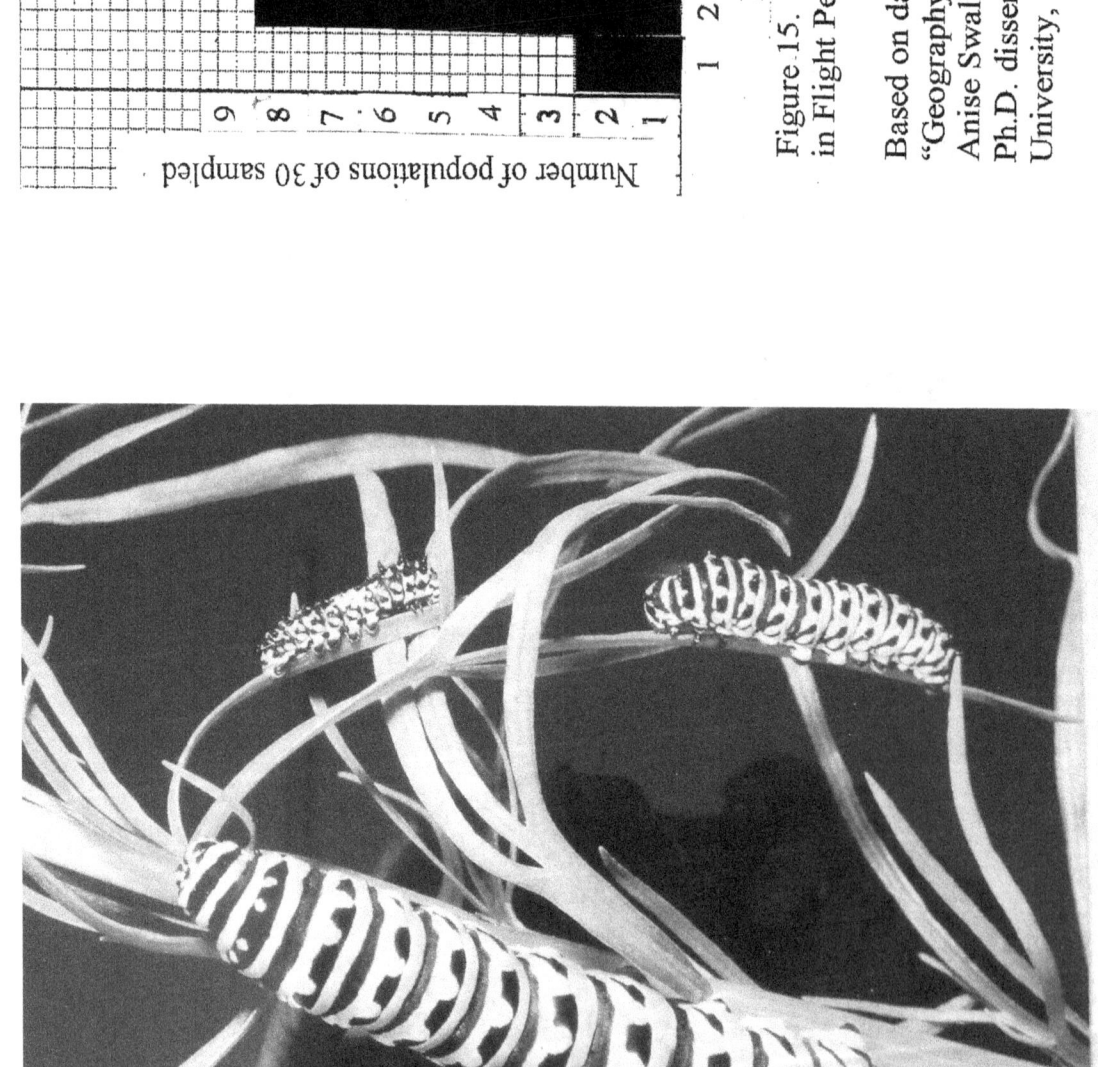

Number of populations of 30 sampled

Months of flight

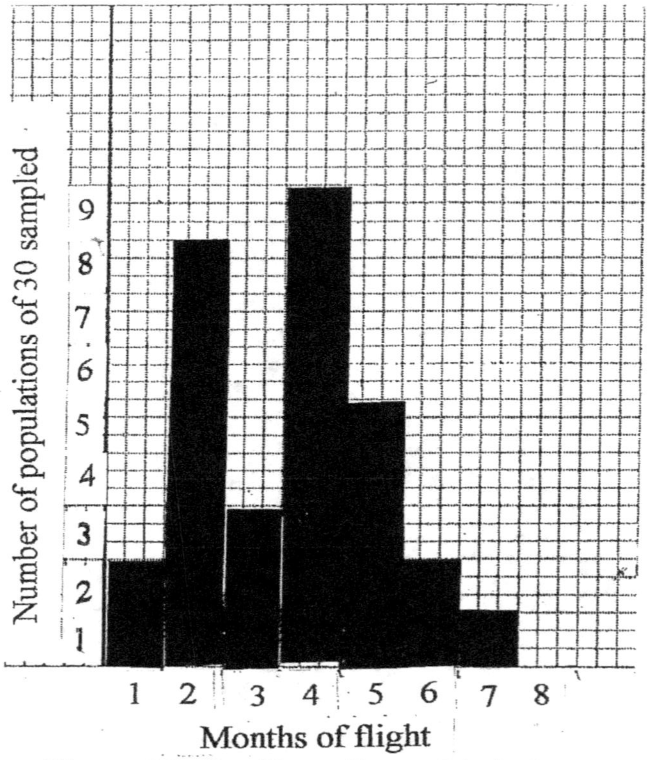

Figure 15. Papilio zelicaon Variation
in Flight Period Duration

Based on data from Wayne F. Wehling:
"Geography of Host Use, etc. in the
Anise Swallowtail, P. zelicaon",
Ph.D. dissertation, Washington State
University, 1994, figure 4.

Figure 16. Three instars of
Papilio bardii oregonius.
Shown on Artemisia dracunculus

Figure 17. Early instar larvae compared
a. Papilio bardii oregonius
b. Papilio indra
c. Papilio zelicaon

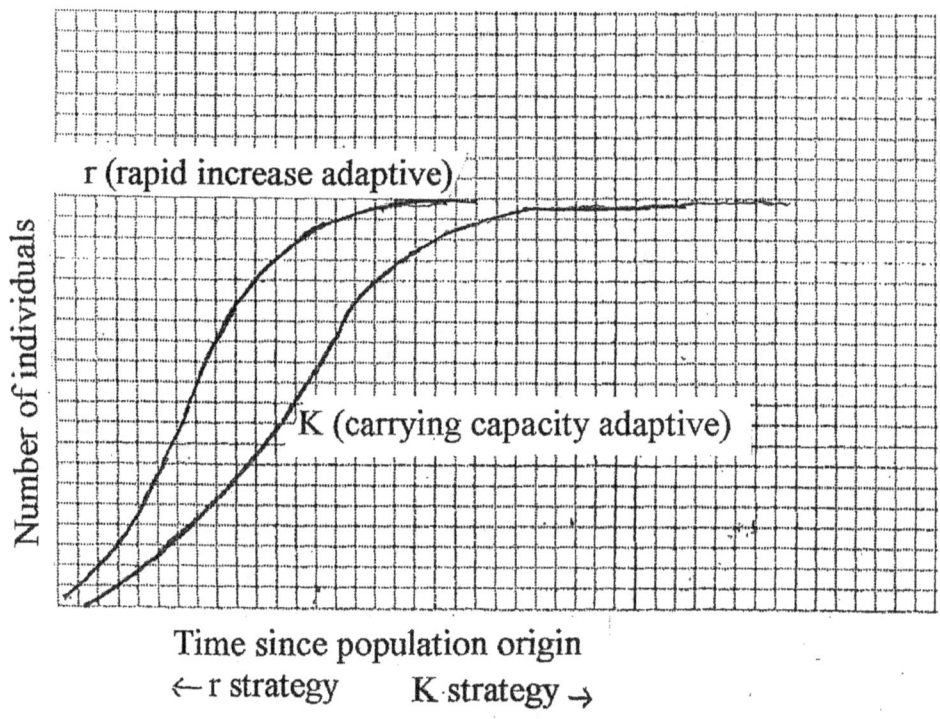

Figure 18. Typical Population Growth Curves

Figure 19: Cliffs of Remote Lemhi Mountains Valley, Idaho

Endnotes

1 Stephen Hawking. *A Brief History of Time*. 10th anniversary edition. New York: Bantam,1998, 190.

2 The meaning here is a somethingness transcending the relative somethingness of our universe and of which such relative somethingness is derived.

3 If absolute somethingness must exist without an origin, why should not its aspects, such as infinite complexity, be a part of this primary existence? As seems apparent, the relative somethingness we experience evolves its complexity. But perhaps this, in a sense, is an illusion of our limited perspective. Could it not be that our "existence potential" is a primary reality which then unfolds within our relative somethingness arena? (Another wording is "our potential exists in God's mind and is actualized by his creative acts.") From the perspective that God has no origin and is immutable, so must our existence potential have no beginning, and thus we too are a part of this primary complexity of infinite somethingness. Further, if so, how can our existence potential ever terminate?

4 However, in Steinhardt and Turok's ekpyrotic universe model, negative gravitational energy is held to be "infinite." And if such "somethingness" is recognized as transcendent (as well as immanent) to our universe, then indeed, our universe would constitute a finite, relative somethingness. (See Steinhardt and Turok[0], Endless Universe, [New York: Doubleday, 2007], 191-192.) Further, apparently distribution within dimensions allows the existence of relative somethingnesses and avoids the "why not everything (coincidentally)?" dilemma.

5 Perhaps such a lack of boundaries would not exclude circularity such as the symbol for infinity (∞) suggests.

6 Even so, one might accept by "faith" that absolute somethingness is finite and therefore does not include infinite sentience, God. However, in addition to being illogical as we have argued, this perspective encounters difficulties with the anthropic principle. If there is neither an infinity of universes (infinite somethingness), nor God, the designer, then what accounts for the large number of parameters that are precisely right in our universe allowing our existence? That absolute nothingness does exist is a perspective elaborated in Lawrence Krauss's book, *A Universe from Nothing. Why there is Something Rather than Nothing.* (New York: Free Press, 2012). See Appendix III for a discussion on the issue.

7 A human inhabitant of an island leaves a footprint in the seashore sand by distorting and compressing the sand so that it conforms to the presence of his or her foot structure, heel-to-toe. The created impression thus is reflective of the actual foot but obviously is not the foot itself. What then is the analogous effect of the presence of infinite sentience upon somethingness such as the relative somethingness that constitutes our universe? Would it not be the imposition of order? Order that is reflective of the infinity of this sentience and hence, comprehensive within our universe.

8 In considering "gaps" in the intrinsic order of our universe, we must distinguish between actual gaps, such as quantum uncertainty represents, and gaps that are, in fact, only in our knowledge of nature, not in nature per se.

9 Quantum entities (etc.) have probability waves requiring coherence and implying a multiplicity of at least potential realities, at least according to the Copenhagen interpretation followed here. A contemporary concept in quantum mechanics holds that these waves may decohere under the influence of aspects of their environment, and thus collapse into a particular reality. The Coherent Quantum Domain perspective suggests that the "probability" wave consists of "energy" spread over the entire wave until it converges into a "reality point" which constitutes its collapse. But before decoherence, this reality is uncertain, and thus the future is indeterminate or open. Apparently, which realities result is of unknown cause, in terms of natural law. And if causes were not orderly, then superquantum disorder should be the result. Yet the

order of classical physics emerges. Some hold that averaging of the multiplicity of quantum collapses yields this order. Why such "averaging" would happen to generate and sustain such order without a designer seems remarkable to at least this biologist. But then there is the "many worlds" interpretation, which I consider the least parsimonious. (e.g., see M. Kaku, Parallel Worlds [New York: Anchor Books, 2006].) This perspective holds that each possibility is real, but in a separate universe. Here decoherence is considered the splitting off of what were coherent universes, each in response to a quantum event. But wherein is the probability of such a wave if each "possibility" rather is actual, even if in a separate universe?

   In his book, The Hidden Reality, (New York: Alfred A. Knopf, 2011), on page 237, Brian Greene leaves open the issue as to which perspective of probability wave function is actual. His "something else entirely" (last paragraph) is what I suggest. That is that this may, indeed, be God's "causal joint," at least in part.

[10] David Bartholomew suggests that quantum probability waves are actually our state of knowledge prior to observation and that our act of observing is really an act of sampling such as may be done in any statistical process. Thus, he suggests, our account of the quantum ("micro") level processes are no better than a (standard) probability representation. (David Bartholomew, God, Chance and Purpose [New York: Cambridge Univ. Press, 2008],149.) However, this hardly seems to account for the "quantum weirdness" of which quantum physicists expound. For instance, do standard probability curves evolve as described by Schrodinger's equation? And what of the perspective that it is interaction with the quantum "environment" that actually results in decoherence and collapse, not just our intentional measurments, a perspective held to by Brian Greene and others. (e.g., see B. Greene, The Fabric of the Cosmos [New York: Alfred A. Knopf, 2004], 209-213.)

[11] Brian Greene, ibid.,207, suggests that a large object's wave function is a composite of those of the many quantum units of which it is composed. Sean Carroll, in his book From Eternity to Here (New York: Dutton, 2010), suggests that the whole universe has a succession of eigenstates, that is realities resulting from the coordinated collapse of all the constituent probability waves, and that each eigenstate is irreversible. Apparently, there cannot be simultaneous eigenstates of position and momentum. Carroll, ibid., 243, holds that successive position and momentum eigenstates are separated by wave function status. But what determines the eigenstate, out of all the possibilities, that actualizes at each time horizon? Greene, ibid., 212, recognizes this issue when he asks "how does one outcome win and where do the other possibilities go?" Carroll, ibid., 239, poses the question "is consciousness playing a crucial role in the fundamental laws of physics?" The biblical theist has an obvious answer! Colossians 1:17, the Bible, states "He is before all things and in him all things hold together." Thus, God in his omnipotence and omnipresence (immanence) sustains order throughout the universe. And thus clasic level reality remains consistent. Again, can such superquantum order really be achieved merely by the averaging out of quantum event collapses or may it rather be dependent upon God's action in his immanancy as considered above?

[12] A quotation from the Bible is pertinent here. Mark 10:18 and also Luke 18:19 state: "And Jesus said unto him, why do you call me good? No one is good except God alone."

[13] Ecologists say the population has exceeded its carrying capacity. We humans have raised our carrying capacity level by our technology, but at the expense of other species' resources. Even so, all earthly resources are finite.

[14] Suppose one were to define good versus bad as a degree of entropy, complete entropy being absolute badness. If, then, somethingness is infinite and eternal, how is it that entropy is not complete and final? And is that it is not, not evidence of the existence of goodness, and thus of the "good" God of the Bible,Matthew 19:17, Mark 10:18, Luke 18:19? But why then would a good God allow entropy (bad?) at all? We see, however, that the entropic process characteristic of at least this universe, allows the creation of forms of complexity. Thus, some "bad" (entropy) is actually good and God's good character is not compromised. Carroll, ibid., chapter fifteen, deals with the entropy issue. It is evident that the Big Bang model has problems accounting for its initial low level.

[15] Hence, the significance of God's "commandments." (For instance, see Romans 5:14 and 5:18-19.)

[16] In considering rainstorm "waste," we tend to overlook the reality that the very water in our hose is made available by God's "random" rainstorms.

[17] James Gleick. *Chaos: Making a New Science* (New York: Penguin Books, 1998).

[18] F. F. Bruce, ed. The International Bible Commentary (Grand Rapids, Michigan: Zondervan Publishing House (Marshall Pickering), 1986), 848.

[19] Google "Enuma Elish" for more on this subject.

[20] But why can't these laws just "be"? What requires God's immanency here? These questions take us back to our earlier considerations of why anything bothers to exist at all and the reasoning that somethingness must be infinite, including infinite complexity, God, who would then be involved in all somethingness. (Infinite complexity must involve all levels of complexity and thus all somethingness.)

[21] Steinhardt and Turok, *Endless Universe: Beyond the Big Bang* (New York: Doubleday, 2007).

[22] Stuart Kauffman, *Beyond Reductionism: Reinventing the Sacred*. New York, Philadelphia: Perseus Books Grp., 2010. Here Kauffman holds quantum events to be "without cause" (natural or otherwise). Such a conclusion is obviously speculative. Apparently, it is the basic premise for his perspective that at least biotic complexity is emergent through natural process alone. Thus, it can be argued that that perspective is no more scientific (empirically testable) than that of theistic determinism as held here.

[23] Of course "the God," as here conceived, does not have gender. I use the term "himself" only as an analogy. ("Herself" would be as fitting except that it gives conflict with Mary as Jesus' mother and God as Jesus' father. "Itself" hardly fits either. Perhaps what is needed is a more accurate pronoun in the English language.)

[24] Hugh Ross, *The Creator and the Cosmos* 3rd ed. (Colorado Springs: Navpress, 2001), 110.

[25] In physics, plasma is a more energized state beyond a gas (e.g., see Brian Clegg, *Before the Big Bang* [New York: St. Martin's Press, 2009],107).

[26] Note that even the Enuma Elish creation account of the Babylonians, referred to earlier, recognizes rain originating in clouds, although this account, too, has "waters" (urine?) restricted above in "heaven" (Tiamat's frontal body half).

[27] Note that the biblical statement has the plants growing from the "dry land" and aquatic animals living in the separated "waters." Thus, our suggestion that the "dry land" is allegorical for "matter" and the "waters" for "plasma" distinguishes it from the writer's intended meaning. The suggested allegory certainly is "unwitting."

[28] A gene pool is the combined genetic information contained in the members of an interbreeding population.

[29] This is because recessive alleles are not expressed in diploid individuals that are heterozygus for the trait, in which the recessive allele is masked by the expression of a dominant allele of the given gene. Sexual reproduction promotes gene "shuffling" and recombination-enhancing variation in phenotypes.

[30] For instance, one in every four offspring will be homozygous recessive if both parents are heterozygotes. (Only in the homozygous condition is the recessive allele expressed in the phenotype, assuming complete dominance of the opposing allele.) Two of every four, however, retain the recessive gene allele unexpressed and thus serve as the bearers of the gene pool reserve.

[31] We note that a species' role in the dynamics of ecology is not only to take from the system, but to give as well. Predators take their share, then scavengers and decomposers depend upon the remains. Thus, a given species yields its "less fit" (as well as its aged). But, in so doing, the species long-term good is served: its gene pool is refined and resource is freed for a new generation. Indeed, the implication of Jesus' words regarding fallen sparrows (Matt. 10:29) are profound in their allegorical context.

[32] For example, see Alexander V. Badyaev, "The Beak of the Finch: Coevolution of Genetic Covarience Structure and Developmental Modularity During Adaptive Evolution," *Philosophical Transactions of the Royal Society* 365 no.1543, (April 2010), 1111-1126.

[33] Some groups, such as certain aphids and many plants, have both "parthenogenetic" phases and bisexual reproductive phases giving them the advantages of both systems.

[34] Adam would have had difficulty with these unless he had a microscope or some equivalent! (Genesis 2:19)

[35] The ancients' concept of "Earth" was not that of a spherical planet. But we have concluded that that must have been what God meant as the author of their inspiration. Or might God actually have meant what we now conceive of as the "earth brane" as indicated in our previous discussion?

[36] Luke 3:38: Here Adam is identified as a "son of God."

[37] Note that the Bible states that even creatures that die in the sea "return to the dust," Psalm 104:29. This would suggest that biblical "dust" must mean something other than dry, powdered debris.

[38] Probability waves do have peaks (although their positioning is apparently dynamic), implying that certain realities are more likely to occur than others. However, even the shape of a probability wave with its peak's position must somehow be determined. May it be that even Schrodinger's equation is an expression of God's ongoing creative and sustaining action in nature?

[39] Sean Carroll in *From Eternity to Here*, [(New York: Dutton, 2010), 239-240] states, "surely we don't want to suggest that the phenomenon of consciousness is somehow playing a crucial role in the fundamental laws of physics?" The perspective of the present thesis is that, indeed, it is! Infinitely sentient (conscious), "panimmanent" God somehow is entangled within the quantum realm and is actualizing his will to maintain order and determine outcomes. Theistic Determinism!Science, as a practice of us humans, deals with the aspects of reality that we can study in our limited context. To conclude that this physical context is the extent of all reality certainly would take faith. The perspective of divine determinism may venture beyond our "science," but that its implications provide meaning to (are concordant with) science certainly evidences reality beyond the purview limitations of that science.

[40] As a consideration of human geneologies establishes, even family lines involve time-lines, lines that intertwine in sexuallly reproducing species. This intertwining establishes the population as a unit with its own timeline.

[41] John Feltwell, *The Encyclopedia of Butterflies* (New York,etc: Prentice Hall General Reference, 1993), 52.

[42] Felix Sperling and Paul Feeney in Scriber, Tsubaki and Lederhouse editors, "*Swallowtail Butterflies*: *Their Ecology and Evolutionary Biology*," Scientific Publishers, (Gainsville: 1995): 303.

[43] *Papilio machaon* vs. *P. bairdii* = .2% divergence, *P. machaon* vs. *P. zelicaon* = 1.25% (2.5 mya / 2) divergence based upon mDNA comparisons (Sperling and Feeney, ibid., 303-304).

[44] Interesting examples of this involve multiple-year salmon runs and certain butterfly populations with two-year life cycles. For instance, the pink salmon *Oncorhynchus gorbuscha* in the Pacific Northwest has a strict two-year life cycle. If every year were to have a run, each run would be essentially genetically identical, but reproductively isolated and thus behave as a distinct species. Each run would have the same ecological niche, although simultaneously occupying it at different life stages. Suppose that in a given year the run, then in its second year, were to overwhelm its food resources, so that the smaller fish of the run in its first year were starved out. The result should be the survival of only that one of the alternate year runs. And such survival is the case in nature. Other salmon species such as silver salmon, *O. kisutch*, and chinook salmon, *O. tshawytscha* have apparently avoided this outcome by having runs that take more than two years to mature, with fish that make spawning runs out of phase of their usual year cycle. Males that do this are known as "jack salmon." Thus, there is gene flow between the runs keeping them functionally as a single species. For apparently similar reasons, certain buterfly species with obligate two-year cycles fly only alternate years, for example, far north species such as *Boloria astarte*.

[45] Considering the status of his contemporary science, could the Psalmist really have known the profundity of his poetic account of God's action in nature? Perhaps we again have come upon evidence of inspiration from transcendent sentience, God.

[46] Parasitoids are distinguished from parasites in that they eventually kill their hosts, whereas parasites only use a non-lethal amount of sustenance, at their host's expense.

[47] Keiichi Honda, "Larval Osmeterial Secretions of the Swallowtails (Papilio)," *Journal of Chemical Ecology*, 7, no. 6(1981).

[48] *Papilio indra* is likely a daughter species derived from *P. zelicaon*, possibly as a reproductive propagule in southern California where its present subspecies, *pergamus*, is presently found. Here young larvae retain a white-patch marking and may rest on the upper leaf surface. Because this is the usual habit in machaon complex species in general, this is taken as evidence of pergamus's primitive condition. How pergamus relates ecologically to local zelicaon larvae is an account beyond our context and apparently not yet investigated.

[49] To minimize my sampling impact on these study populations, I not only sampled in alternate years, but also divided the host plant stand into thirds. The middle third was left untouched. The end thirds were alternately sampled in the sequence of field study years.

[50] In some cases, high reproductive potential has other functions than "r adaptation."For instance, sea urchin females may produce many thousands of small eggs each time they spawn, usually annually. When fertilized by the clouds of sperm the males release, the resulting small larvae, known as the echinopleuteus stage, are widely dispersed by ocean currents so that any vacant rock surface is likely to be encountered by at least a few larvae, which then attach and become comparatively sedentary. The vast majority of the spawn and larvae produced become a bonanza for filter feeders such as sponges and barnacles.

[51] For instance, see ;ltanthroplogy.ua.edu/bindon/ant570/topics/Modernhumans.pdf;mt[0].

[52] Note that, like our octopus, the chimp had learned to launch "ammunition" at irritating intruders.

[53] Curtis Marean (Ibid., 59) states that he has good evidence that the Pinnacle Point people used fire-generated heat to treat a type of rock (silerete) so as to make it more workable in producing primitive tools. This technology was apparentlly well-established 72,000 years ago and likely begun as far back as 164,000 years ago.

[54] If a ruler has all he or she desires of personal possesions, then presumably they will rule impartially in the distribution of further resource. However, if the system unduly favors an upper class of its members at the expense of a lower class, there may be revolt, such as that of the French Revolution and the recent revolts in the Middle East. The most stable system would be that in which all members have equal socio-economic opportunity. Jesus admonished this principle in his statement, "love your neighbor as you love yourself," (Matt.19:19, The Bible). Note that this would solve the "tragedy of the commons" problem.

[55] In a 1985 Brookings Institute study report titled, "Religion in American Public Life," A. James Reichley et al. make the following conclusions. Representative "government depends for its health on values that over the not-so-long run must come from religion." "Banishment of religion does not represent neutrality between religion and secularism; conduct of public institutions without any regard to religion is secularism . . ." Without religion to provide a transcendent moral base, "either the self or society must finally be regarded as sovereign." Thus, there would be no effective basis for individual rights.

[56] S. S. Madar, *Biology* 10th ed. (New York: McGraw Hill, 2010), 295.

[57] I once had the occasion of at least sensing the role of being potentially cannabalized. My wife and my sister-in-law were on vacation with me in Fiji. We had been instructed by the tourist agency to be sensitive to the native Fijians' history of cannabalism and to not call attention to their guilt of that practice. But this turned out to be a "set-up." In fact, a rather effective one for us. We had occasion to be on a tour of a typical village. Our guide was a Fijian woman in her mid-thirties. After showing us a good portion of the village and explaining the significance of such things as its architecture and design, she stated that she would tell us of her peoples' history, especially as it related to outsiders. She began with these words, which have remained prominent in my memory. "Many years ago, the missionaries came to Fiji and we ate them." As she patted me on my shoulder, she continued. "We found that white people taste very good!" Fortunately for me, her implications for my fate were in jest. But I think that she took pleasure in my rather shocked response.

[58] In this context there is pertinence to the admonishments given to "slaves" in Ephesians 6:5.

[59] Robert E. L. Faris deals with especially "primary group" formation and function in his book, *Social Psychology*(New York: The Ronald Press Co.,) 1952. See especially "The Tendency to Form Groups" on page 255.

[60] What is the adaptive advantage of non-kin, true altruism in a social system where there is no moral base?

[61] One notes the difference in social dynamics in a shower/locker room compared to the dynamics in, say, a conference room where the same people are fully clothed. However, it may be that nudist colonies function adaquately. I leave that observation to others.

[62] At a later date, this person has undertaken weight-loss treatment and is in training for a home-based, on-line opportunity. And we, as neighbors, yet have opportunity to exercise our love.

[63] My father, Addison L. McCorkle, had the middle name "Lawrence." Also, he had red hair as apparently did "Lawrence of Arabia." Hence, he was occasionally mistaken for this famous person. I think of him now as "Lawrence of Iraq," certainly a hero to me.

[64] I often wonder what has become of the small Christian community my parents and the others in their mission established there in Iraq, especially now, in times of persecution. During later years, while living in the Seattle, Washington area, my father was able to occasionally befriend Iraqi students attending the University of Washington. I especially remember one young man, an anthropology major, who visited our home on several occasions.

[65] Indeed, if the multiverse hypothesis is actual, then what determined in which of the universes I avoided calamity? Again, I argue that God is in control. (He well could actualize my survival in all of them, so why the need for duplication of universes?)

[66] Sean Carroll *From Eternity to Here: The Quest for the Ultimate Theory of Time* (New York: Dutton, 2010), 245.

[67] And may it be that this "awareness" constitutes our "soul" which would be restricted to our earthly body until its death, and at that point received into the heavenly avatar-our spirit body?

[68] The biblical account of heaven apparently does not elaborate on its "astronomical" features. Are there "galaxies" with stars and planets? Could it be that each such astronomical body in this universe, the Earth brane, has a counterpart heavenly "avatar" in the heaven brane? And if, indeed, dark matter is actually heaven brane matter, detectable by us only by its gravity (closed-ring gravitons), then we should expect to locate dark matter concentrated in the vicinity of the astronomical bodies of our universe. Apparently, this is the case for the most part. (See also "mirror particles" and "parity symmetry" e.g., Google.) And would heaven (paradise), then, be on a planet avatar of Earth? Here we can only speculate.

[69] Carroll, Ibid., 280-281.

[70] Carroll (Ibid., 241) holds that past eigenstate information is destroyed by subsequent reversion to wave function status. But he also states on page 274, "the laws of physics preserve the information needed to specify a state as the universe evolves from moment to moment." Here I argue that, in light of the potential of Heisenberg uncertainty, it is rather the consistency resulting from theistic determinism that constitutes the laws of physics and thus the orderliness of subsequent eigenstates. (Collosians 1:17. "and in Him all things hold togther."NIV.)

[71] See Carroll, Ibid., 280-281.

[72] Presumably, each needed particle would thereby disappear from its former location to reappear at the resurrection "construction" site. One might speculate that one's original body molecules would reassemble. But what, then, of shared molecules such as would be the result of e.g., cannibalism?

[73] I have not chosen to speculate on hell in this thesis. The Bible speaks of the final fate of those who willfully choose to disobey God's will as being cast into a "lake of fire" (Revelation 20:15).

www.ingramcontent.com/pod-product-compliance
Lightning Source LLC
Chambersburg PA
CBHW081222280526
45787CB00006B/2491